SCIENTIFIC REALISM

SCIENTIFIC REALISM

Edited, with an Introduction, by

JARRETT LEPLIN

UNIVERSITY OF CALIFORNIA PRESS
Berkeley   Los Angeles   London

University of California Press
Berkeley and Los Angeles, California

University of California Press, Ltd.
London, England

Copyright © 1984 by The Regents of the University of California

Library of Congress Catalog Card Number: 84-40311
ISBN 0-520-05155-6 cloth
     0-520-05326-5 paper

Printed in the United States of America

1 2 3 4 5 6 7 8 9

Contents

Preface

This book originated in a conference on scientific realism sponsored by the Department of Philosophy of the University of North Carolina, Greensboro, March 26-28, 1982. The first five essays in this book are the principal papers from that conference. Commenting on these papers were, respectively, Larry Laudan, William Boos, John King, Roger Jones, and Jarrett Leplin. Commentaries are not included since they were taken into account in preparing the final versions of the essays.

The remaining six essays were selected to represent the major current positions and directions of research on the topic of scientific realism. Represented only indirectly are such seminal writers on realism as Thomas Kuhn and Paul Feyerabend, whose works are well known and widely available.

Six of the essays have been published previously; Ronald Laymon's in *PSA* 1982; Hilary Putnam's as lecture 2 of *Meaning and the Moral Sciences* (London: Routledge and Kegan Paul, 1978); Ian Hacking's in *Philosophical Topics*, vol. 13, no. 1; Jarrett Leplin's in *Studies in History and Philosophy of Science*, vol. 12, no. 1; Larry Laudan's in *Philosophy of Science*, vol. 48, no. 1; and Bas van Fraassen's in *Journal of Philosophy*, vol. 73, no. 18.

I wish to thank the participants in the Greensboro Symposium, Ernan McMullin especially, for cooperation and assistance. I also wish to thank the University of North Carolina at Greensboro for its financial support of the Annual Symposium in Philosophy.

J. L.

1

Introduction

Jarrett Leplin

Hilary Putnam seems to have inaugurated a new era of interest in realism with his declaration that realism is the only philosophy that does not make the success of science a miracle. None of the authors of the present papers either denies that science is successful or holds the success of science to transcend human comprehension. But there is much disagreement as to what that success consists in, how it is to be explained, and the role of realism in its explanation. Like the Equal Rights Movement, scientific realism is a majority position whose advocates are so divided as to appear a minority. The following theses are characteristic realist claims no majority of which, even subjected to reasonable qualification, is likely to be endorsed by any avowed realist:

1. The best current scientific theories are at least approximately true.
2. The central terms of the best current theories are genuinely referential.
3. The approximate truth of a scientific theory is sufficient explanation of its predictive success.
4. The (approximate) truth of a scientific theory is the only possible explanation of its predictive success.
5. A scientific theory may be approximately true even if referentially unsuccessful.
6. The history of at least the mature sciences shows progressive approximation to a true account of the physical world.

7. The theoretical claims of scientific theories are to be read literally, and so read are definitively true or false.
8. Scientific theories make genuine, existential claims.
9. The predictive success of a theory is evidence for the referential success of its central terms.
10. Science aims at a literally true account of the physical world, and its success is to be reckoned by its progress toward achieving this aim.

What realists do share in common are the convictions that scientific change is, on balance, progressive and that science makes possible knowledge of the world beyond its accessible, empirical manifestations. Unless progress is understood in purely pragmatic terms and knowledge is held not to require truth by correspondence, antirealists will reject these convictions. The objections they offer center on two principal problems.

One problem is historical. Whatever continuity may be discerned in the growth of empirical knowledge, theoretical science has been radically discontinuous. Scientific views about the ultimate structure and lawlike organization of the world have frequently been overthrown and replaced by incompatible views. Much of this discarded science was, for an appreciable time, eminently successful by the standards we employ in assessing current science. The inference seems inescapable that the evidence available to support current science is by nature unreliable and systematically underdetermines what ought to be believed about the world beyond our experiences. Scientific theories, however well secured by observation and experiment, are inevitably fallible. Nor is there any basis for expecting the future evolution of scientific standards and methods to provide a more secure foundation for scientific knowledge. For methodological developments that have occurred thus far, whatever improvements they have generated at the level of human interaction with nature, have failed utterly to resolve the basic dilemma of the underdetermination of theory.

Theory change alters the characterization science gives of the entities and processes alleged to constitute the world. Even where the same entities and processes appear to be countenanced by successive theories, their descriptions are so altered as to make it impossible to discern referential stability if reference is at all dependent on accounts of the nature of the referent. Thus, history appears to discredit our ability to identify the actual constituents of the world as much as it does our ability to learn their true natures.

The second problem concerns the explanation which an imputation of truth or approximate truth to a theory can give of its empirical success. Even if a theory were true or approximately so, that fact about it could easily fail to be reflected in success. A true theory, unless complete in some global sense, might be too remote from our experience to affect it in any way, or to affect it differently from some false alternative. Inaccuracies in the background assumptions made in applying a theory might produce predictive failure. And the retreat to approximate truth, in addition to the vagueness it introduces, forfeits even a presumption of success should the area of inaccuracy happen to be crucial to our experience of the world. A theory that gets most everything right, missing just some fact about photons, say, might easily number among the least successful in laboratory appraisal.

Conversely, a theory not even approximately true could be empirically impressive through the invocation of opportune auxiliary assumptions or chance agreement at the level of testable generalizations with one more verisimilar. As alternative theoretical structures can often be posited for the same phenomena, such agreement should occasion no surprise. And the presence of true statements within the consequence classes of false ones needs no explanation. One may hope that the application of theories in new areas will yield differential success, that false theories will eventually yield a preponderance of false, testable consequences. But whether theories are necessarily discriminable in this way is dubious, and at any given time the evidential picture is indecisive. The successful extension of a theory to new areas yields a greater body of corroborations from which further experience may yet diverge. And the idea that successful extendibility has any special epistemic significance as against the sheer quantity of the resultant successes is difficult to sustain, extendibility reflecting as much on the limitations of our initial perspective as on the merits of our theory.

A further and more fundamental aspect of the alleged connection between truth and success has recently emerged as a source of antirealist argumentation. This aspect concerns the assumption implicit in realist views based on theses 3 and 4 of the legitimacy of abductive reasoning or "inference to the best explanation." If such reasoning is indeed legitimate, it may be used within science to infer the truth of hypotheses directly from their explanatory and predictive successes, thus obviating recourse to the explanatory power of realism with respect to scientific success generally. If, instead, such reasoning is suspect, if explanatory status is judged an insufficient basis for inference, then an explanationist defense of realism can be no more cogent than the suspect support which observational evidence provides theoretical hypotheses within science. In either case, realism gains nothing from its alleged explanatory status; if not superfluous it is question-begging.

This argument, powerful as it may appear, has an important limitation. It is ineffective where it is not abduction itself that is questioned, but the need within science to posit unobserved entities in suitable explanations. Even if it is possible to account nonrealistically for individual successes of individual theories, there may yet be explananda for realism in the overall successfulness of scientific method. The proper target of explanationist realism is not the antirealist who distrusts inference to the best explanation, but the nonrealist who doubts that reference to unobserved entities provides the best explanation of what is observed.

I believe it is fair to say that neither the problem posed by the historical record of theory change nor the problem about the connection between truth and success has been solved even to the satisfaction of realists. At present, the most promising realist strategies are to argue that these problems are indecisive or to argue for realism independently, so that one has, as it were, an existence proof for the solutions one lacks.

Thus, one might follow Putnam in invoking the causal theory of reference on which radical change in accounts of the nature of the referents of scientific terms is compatible with referential stability through theory change. Or one might deny that approximate truth requires referential success. Approximate truth, after all, is a concept in need of analysis; even if the truth of a statement requires that its purportedly referential terms be genuinely referential, it is not clear that approximate truth requires this. Or one might deny the alleged datum of discontinuous conceptual change, insisting on a sufficiently cumulativist reading of history to permit referential stability on a more traditional, Fregean theory of reference.

Alternatively, one might attempt to identify features of scientific method or scientific reasoning that are unintelligible or empty but for realism. The argument then is that no form of antirealism can do justice to the scientific enterprise. This approach must either supply a noncircular defense of abduction, or show in just what respects a realist explanation of science is superior to the explanations which theoretical hypothesis give of the data they successfully predict.

Among the present papers, those of Levin, Glymour, Laymon, Boyd, and Hacking pursue the latter alternative; those of Putnam, McMullin, and Leplin, the former. Levin's concern is to refute instrumentalism by portraying it as a view incapable of yielding distinctions crucial to understanding the content of theories. Thus Levin endorses thesis 8. He finesses the historical problem by endorsing thesis 9, denying that the record of scientific failure supports any inference as to the credibility of current science. He denies, however, that realism gains any support from the explanationist argument reflected in theses 3 and 4. Glymour argues that the comparative assessment of explanations offered within science often requires thesis 2, showing inter alia that extant accounts of scientific explanation fail to sustain comparative assessments even in paradigm cases. Laymon argues that recognition of the role of idealization in theory testing leads to an account of confirmation requiring thesis 9. Boyd's principal concern is to show that certain features of scientific method, in particular the instrumental reliability of the theory-dependent methodology of the "mature sciences," lead inevitably to realism in the form of theses 1, 2, and 4. Thus he argues that neither the empiricist tradition, which invalidates all inference to unobservables on the basis of the underdetermination in principle of theories, nor the constructivist tradition, which denies the independent reality of the objects of scientific knowledge on the basis of the theory dependence of method, has the resources to explain the empirical success of the mature sciences. This focus on the explanatory resources of realism in contrast with that of its major alternatives is supposed, by Boyd, to provide the ultimate rejoinder to antirealist arguments which attack the legitimacy of abductive reasoning as such. Hacking's version of realism is closest, among the options distinguished, to thesis 2. But it purports to differ significantly from familiar approaches in focusing on the nature of experimentation in science as against theories and their successes. Hacking sees experimentation as a largely autonomous activity; when liberated from the presumed constraints of theory testing, it is seen to have realist implications. He supposes that these implications obviate questions about the reliability and underdetermination of theories.

Putnam formulates the basic explanationist case for realism as the only philosophy that accounts for the success of scientific method. He allows that realism requires theoretical continuity across conceptual revolutions; it must be possible to recognize stability of reference, at least to the extent of assigning referents to past theories from the viewpoints of their successors, and to recover past theories as limiting cases of their successors. Should this not be possible, should it turn out that paradigmatically successful theories get replaced by theories postulating radically different entities and laws to which nothing formerly recognized approximates, then realism would not be a tenable position. This would not mean, however, that we would abandon altogether the notions of truth and reference. We would still have important uses for concepts possessing the formal properties of truth and reference, such as are captured by Tarski's theory of truth; only the concepts we would use would be theory relative—such as warranted assertability or provability within a system—rather than classical ideas based on correspondence to theory-independent fact.

Putnam does not, however, believe that the history of science requires such a retreat from the classical ideal. By denying that the descriptions used in fixing the reference of scientific terms are synonymous with them, it becomes possible to preserve reference across the substantial descriptive changes that accompany theory replacement. The key is a principle of charity according to which a historical figure is to be credited with having referred to whatever entity countenanced by current theory answers to enough of the descriptions he used so that it is reasonable to suppose he would have identified this entity as the referent, had he known more of the relevant facts. Of course, "would have identified this entity as the referent" can only mean "would have altered his descriptions," which indeed he would on the supposition that he knows them to be mistaken. So what the principle of charity amounts to is the requirement that some substantive claims of past science survive as a condition for preservation of reference, together with a plea that some is enough.

The papers of McMullin and Leplin attack the historical problem directly. Leplin maintains that the historical record allows for referential stability, and develops an explanationist defense of realism along the lines of theses 3 and 4 which presupposes such stability. Leplin's approach purports to advance beyond the defense embodied in these theses by adducing a variety of independent forms of scientific progress to serve as explananda for realism. McMullin's realism is close to thesis 6, but the progress it diagnoses does not depend on achieving truth. Rather it occurs when a rejected theory is successful in indicating the direction to be taken by its successor. McMullin's analysis of such indications invokes the notion of metaphor. The existential commitments of theories are to be read metaphorically, and it is on the fertility and continuity of metaphor that the progressiveness and referential stability of theory change depend. Formal incompatibilities among theories fail to defeat realism because they do not preclude continuity at the level of metaphor.

The papers of Laudan, van Fraassen, and Fine are avowedly antirealist. Laudan develops the historical problem and the problem about truth and success in great detail, arguing that the realism of theses 1, 2, 3, 6, and 9 is empirically refuted. He attempts to convict the realist of violating, in his philosophical interpretation of science, standards of evidence he would certainly impose on science itself. This, in effect, is a version of the antiexplanationist argument described above: if hypotheses are not confirmed by the evidence they are introduced to explain, neither is realism confirmed by the success it purports to explain; if hypotheses are thus confirmed, realism is unnecessary. Leplin's attempt to distinguish independent forms of progress is motivated by just this problem. Van Fraassen focuses on the problem of underdetermination, arguing for the possibility of empirical equivalence in principle of theories differing in their ontological commitments. Van Fraassen rejects thesis 10, insisting that the aim of science is the empirical adequacy of theories. Fine, proclaiming the death of realism, offers the most sweeping and original attack on explanationism as a defense of realism. According to Fine, the circularity of explanationism could be broken only by employing in defense of realism a form of reasoning more stringent than the abductive reasoning found wanting within science. But no such additional stringency is available, as the contents of individual theories cannot be compared with theory-independent facts to provide the basis for an inductive inference to realism. Fine does, however, sanction the inference from evidence to hypothesis within science, which is very close to realism in the form of thesis 9. The difference is that acceptance of a hypothesis need not, for Fine, involve any of the metaphysical implications which the truth of the hypothesis has been thought to carry in the realist tradition. What Fine calls the "natural ontological attitude" views the explanatory and inferential structures of scientific reasoning as autonomous; metaphysical interpretation is a dispensable superaddition. Thus, Fine agrees with Levin that philosophical accounts of truth and reference have no explanatory role in understanding scientific success or underwriting scientific conclusions. Science all by itself says all that needs to be said or can defensibly be said in response to philosophical questions about the nature and status of scientific knowledge.

As my contribution to this volume indicates, I number among the realists. The qualification is that I see serious historical problems in the way of crediting extant science with the sort of success which it is possible to argue that realism alone can explain. Realism is among the growing number of philosophical theories which like many scientific theories are partly metaphysical and partly empirical; it has implications beyond experience but is testable by experience. Most parties to the dispute tend to suppose that insofar as realism is an empirical thesis, the facts needed to assess it are in. Realism is either warranted by the impressive record of scientific success, or refuted by the discontinuities of theory change or the substantive findings of quantum mechanics. If problems remain in the way of assessing realism, they are not of a kind to be solved by further evidence. Despite this attitude, there is controversy over the nature and interpretation of the evidence as well as over the doctrine itself. We will need more history as well as more philosophy to settle the issues aired in this volume and to reach a viable theory of the nature and scope of scientific knowledge.

2

A Case for Scientific Realism

Ernan McMullin

When Galileo argued that the familiar patterns of light and shade on the face of the full moon could best be accounted for by supposing the moon to possess mountains and seas like those of earth, he was employing a joint mode of inference and explanation that was by no means new to natural science but which since then has come to be recognized as central to scientific explanation. In a retroduction, the scientist proposes a model whose properties allow it to account for the phenomena singled out for explanation. Appraisal of the model is a complex affair, involving criteria such as coherence and fertility, as well as adequacy in accounting for the data. The theoretical constructs employed in the model may be of a kind already familiar (such as "mountain" and "sea" in Galileo's moon model) or they may be created by the scientist specifically for the case at hand (such as "galaxy," "gene," or "molecule").

Does a successful retroduction permit an inference to the existence of the entities postulated in the model? The instincts of the working scientist are to respond with a strong affirmative. Galaxies, genes, and molecules exist (he would say) in the straightforward sense in which the mountains and seas of the earth exist. The immense and continuing success of the retroductions employing these constructs is (in the scientist's eyes) a sufficient testimony to this. Scientists are likely to treat with incredulity the suggestion that constructs such as these are no more than convenient ways of organizing the data obtained from sophisticated instruments, or that their enduring success ought not lead us to believe that the world actually contains entities corresponding to them. The near-invincible belief of scientists is that we come to discover more and more of the entities of which the world is composed through the constructs around which scientific theory is built.[1]

But how reliable *is* this belief? And how is it to be formulated? This is the issue of scientific realism that has once again come to be vigorously debated among philosophers, after a period of relative neglect. The "Kuhnian revolution" in the philosophy of science has had two quite opposite effects in this regard. On the one hand, the new emphasis on the *practice* of science as the proper basis for the philosophy of science led to a more sensitive appreciation of the role played by theoretical constructs in guiding and defining the work of science. The restrictive empiricism of the logical positivists had earlier shown itself in their repeated attempts to "reduce" theoretical terms to the safer language of observation. The abandonment of this program was due not so much to the failure of the reduction techniques as to a growing realization that theoretical terms have a distinctive and indispensable part to play in science.[2] It was only a step from this realization to an acknowledgment that these terms carry with them an ontology, though admittedly an incomplete and tentative one. For a time, it seemed as though realism was coming into its own again.

But there were also new influences in the opposite direction. The focus of attention in the philosophy of science was now on scientific change rather than on the traditional topic of justification, and so the instability of scientific concepts became a problem with which the realist had to wrestle. For the first time, philosophers of language were joining the fray, and puzzles about truth and reference began to build into another challenge for realism. And so antirealism has reemerged, this time, however, much more sophisticated than it was in its earlier positivist dress.

When I say 'antirealism', I make it sound like a single coherent position. But of course, antirealism is at least as far from a single coherent position as realism itself is. Though my concern is to construct a case for realism, it will be helpful first to survey the sources and varieties of antirealism. I will comment on these as I go, noting ambiguities and occasional misunderstandings. This will help to clarify the sort of scientific realism that in the end can be defended.

SOURCES OF ANTIREALISM: SCIENCE

CLASSICAL MECHANICS

It is important to recall that scientists themselves have often been dubious about some of their own theoretical constructs, not because of some general antirealist sentiment, but because of some special features of the particular constructs themselves. Such constructs may seem like extra baggage—additional interpretations imposed on the theories themselves—much as the crystalline spheres seemed to many of the astronomers of the period between Ptolemy and Copernicus. Or it may be very difficult to characterize them in a consistent way, a problem that frequently bedeviled the proponents of ethers and fluids in nineteenth-century mechanics.

The most striking example of this sort of hesitation is surely that of Newton in regard to his primary explanatory construct, *attraction*. Despite the success of the mechanics of the *Principia*, Newton was never comfortable with the implications of the notion of attraction and the more general notion of force. Part of his uneasiness stemmed from his theology; he could not conceive that matter might of itself be active and thus in some sense independent of God's directing power. The apparent implication of action at a distance also distressed him. But then, how were these forces to be understood ontologically? *Where* are they, in what do they reside, and does the postulating of an inverse-square law of force between sun and planet say anything more than that each tends to move in a certain way in the proximity of the other?

The Cartesians, Leibniz, and later Berkeley, charged that the new mechanics did not really *explain* motion, since its central notion, *force*, could not be given an acceptable interpretation. Newton was sensitive to this charge and, in the decades following the publication of *Principia*, kept trying to find an ontology that might satisfy his critics.[3] He tried "active principles" that would somehow operate outside bodies. He even tried to reintroduce an ether with an extraordinary combination of properties—this despite his convincing refutation of mechanical ethers in *Principia*.[4] None of these ideas, however, were satisfactory. There were either problems of coherence and fit (the ether) or of specification (the active principles). After Newton's death, the predictive successes of his mechanics gradually stilled the doubts about the explanatory credentials of its central concept. But these doubts did not entirely vanish; Mach's *Science of Mechanics* (1881) would give them enduring form.

What are the implications of this often-told story for the realist thesis? It might seem that the failure of the attempts to interpret the concept of force in terms of previously familiar causal categories was a failure for realism also, and that the gradual laying aside in mechanics of questions about the underlying ontology was, in effect, an endorsement of antirealism. This would be so, however, only if one were to suppose the realist to be committed to theories that permit interpretation in familiar categories or, at the very least, in categories that are immediately interpretable. Naive realism of *this* sort is, indeed, easily undermined. But this is not the view that scientific realists ordinarily defend, as will be seen.

How should Newton's attempts at "interpretation" be regarded, after the fact? Were they an improper intrusion of 'metaphysics', the sort of thing that science today would bar? The term 'underlying ontology' that I have used might mislead here. A scientist *can* properly attempt to specify the mechanisms that underlie his equations. Newton's ether *might* have worked out; it was a potentially testable hypothesis, prompted by analogies with the basic explanatory paradigm of an earlier mechanical tradition. The metaphor of "active principle" proved a fruitful one; it was the ancestor of the notion of field, which would much later show its worth.[5]

In one of his critiques of "metaphysical realism," Putnam argues that "the whole history of science has been antimetaphysical from the seventeenth century on."[6] Where different "metaphysical" interpretations can be given of the same set of equations (e.g., the action-at-a-distance and the field interpretations of Newtonian gravitation theory), Putnam claims that competent physicists have focused on the equations and have left to philosophers the discussion of which of the empirically equivalent interpretations is "right." But this is not a good reading of the complicated history of Newtonian physics. First and foremost, it does not apply to Newton himself nor to many of his most illustrious successors, such as Faraday and Maxwell.[7]

Scientists have never thought themselves disqualified from pursuing one of a number of physical models that, for the moment, appear empirically equivalent. As metaphors, these models may give rise to quite different lines of inquiry, leading eventually to their empirical separation. Or it may be that one of the alternative models appears undesirable on other grounds than immediate empirical adequacy (as action at a distance did to Newton). If prolonged efforts to separate the models empirically are unsuccessful, or if it comes to be shown that the models are in principle empirically equivalent, scientists will, of course, turn to other matters. But this is not a rejection of realism. It is, rather, an admission that no decision can be made in this case as to what the theory, on a realist reading, commits us to.

What makes mechanics unique (and therefore an improper paradigm for the discussion of realism with regard to the theoretical entities of science generally) is that this kind of barrier occurs so frequently there. This would seem to derive from its status as the "ultimate" natural science, the basic mode of explanation of motions. The realist can afford to be insouciant about his inability to construe, for example, "a force of attraction between sun and earth... [as] responsible for the elliptical shape of the earth's orbit" in ontological terms, as long as he *can* construe astrophysics to give at least tentative warrant to his claim that the sun is a sphere of gas emitting light through a process of nuclear fusion. There was no way for Newton to know that attempts to interpret force in terms of the simple ontological alternatives he posed would ultimately fail, whereas the ontology of "insensible corpuscles," which he proposes in *Opticks*, would prosper. Each of these ventures was "metaphysical" in the sense that no evidence then available could determine the likelihood of its ever becoming an empirically decidable issue. But it is of such ventures that science is made.

QUANTUM MECHANICS

In the debates between realists and antirealists, one claim that antirealists constantly make is that quantum mechanics has decided matters in their favor. In particular, the outcome of the famous controversy involving Bohr and Einstein, leading to the defeat (in most physicists' eyes) of Einstein, is taken to be a defeat for realism also. Once again, I want to show that this inference cannot be directed against the realist position proper.

Was the Copenhagen interpretation of quantum mechanics antirealist in its thrust?[8] Did Bohr's "complementary principle" imply that the theoretical entities of the new mechanics do not license any sort of existence claims about the structures of the world? It would seem not, for Bohr argues that the world is much more complex than classical physics supposed, and that the debate as to whether the basic entities of optics and mechanics are waves or particles cannot be resolved because its terms are inadequate. Bohr believes that the wave picture and the particle picture are *both* applicable, that *both* are needed, each in its own proper context. He is not holding that from his interpretation of quantum mechanics nothing can be inferred about the entities of which the world is composed; quite the reverse. He is arguing that what can be inferred is entirely at odds with what the classical world view would have led one to expect.

Of course, Einstein was a realist in regard to science. But he was also much more than a realist. He maintained a quite specific view about the nature of the world and about its relationship to observation; namely, that dynamic variables have unique real values at all times, that measurement reveals (or should reveal) these values as they exist prior to the measurement, and that there is a deterministic relationship between successive sets of these values. It was this further specification of realism that Bohr disputed.[9]

It is important to note that Einstein *might* have been right here. There is nothing about the nature of science per se that, in retrospect, allows us to say that Bohr *had* to be right. There could well be a world of which

Einstein's version of realism would hold true. And in the 1930s, it was not yet clear that it might not just be our world. We now know that it is not and, furthermore, that this was implicit from the beginning in certain features of the quantum formalism itself, once this formalism was shown to predict correctly. (J. S. Bell's theorem could, in principle, have been proved as easily in 1934 as in 1964; no new empirical results were needed for it.)

What we have discovered as a result of this controversy is, in the first instance, something about the kind of world we live in.[10] The dynamical variables associated with its macro- and microconstituents are measurement-dependent in an unexpected way. (E. Wigner tried to show more specifically that they are *observer*-dependent, in the sense of being affected by the consciousness of the observer, but few have followed him in this direction.) Does the fact that quantum systems are partially indeterminate in this way affect the realist thesis? Not as far as I can see, unless a confusion is first made between scientific realism and the "realism" that is opposed to idealism, and then the measurement-dependence is somehow read as idealist in its implications. It *does* mean, of course, that the quantum formalism is incomplete by the standards of classical mechanics and that a quantum system lacks some kinds of ontological determinacy that classical systems possessed.

This was what Einstein objected to. This was why he sought an "underlying reality" (specifiable ultimately in terms of "hidden parameters" or the like) which would restore determinism of the classical sort. But to search for a completeness of the classical sort was no more "realist" than to maintain (as Bohr did) that the old completeness could never be regained. Recall that realism has to do with the existence-implications of the theoretical entities of successful theories. Einstein's ideal of physics would have the world entirely determinate against the mapping of variables of a broadly Newtonian type; Bohr's would not. The implications for the realist of Bohr's science are, it is true, more difficult to grasp. But why should we have expected the ontology of the microworld to be like that of the macroworld? Newton's third rule of philosophizing (which decreed that the macroworld should resemble the microworld in all essential details) was never more than a pious hope.

ELEMENTARY-PARTICLE PHYSICS

And this dissimilarity of the macrolevel and microlevel is even plainer when one turns from dynamic variables to the entities which these variables characterize. In the plate tectonic model that has had such striking success in recent geology, the continents are postulated to be carried on large plates of rocky material which underlie the continents as well as the oceans and which move very slowly relative to one another. There is no problem as to what an existence-claim means in this case. But problems do arise when we consider such microentities as electrons. For one thing, these are not particles strictly speaking, though custom dies hard and the label 'elementary-particle physics' is still widely used. Electrons do not obey classical (Boltzman) statistics, as the familiar enduring individuals of our middle-sized world do.

The use of namelike terms, such as 'electron', and the apparent causal simplicity of oil-drop or cloud-track experiments, could easily mislead one into supposing that electrons are very small localized individual entities with the standard mechanical properties of mass and momentum. Yet a bound electron might more accurately be thought of as a state of the system in which it is bound than as a separate discriminable entity. It is only because the charge it carries (which is a measure of the proton coupling to the electron) happens to be small that the free electron can be represented as a independent entity. When the coupling strength is greater, as it is between such nuclear entities as protons and neutrons, the matter becomes even more problematic. According to relativistic quantum theory, the forces between these entities are produced by the exchange of mesons. What is meant by 'particle' in this instance reduces to the expression of a force characteristic of a particular field, a far cry from the hard massy points of classical mechanics. And the situation is still more complicated if one turns to the quark hypothesis in quantum field theory. Though quarks are supposed to "constitute" such entities as protons, they cannot be regarded as "constituents" in the ordinary physical sense; that is, they cannot be dissociated nor can they exist in the free state.

The moral is not that elementary-particle physics makes no sort of realist claim, but that the claim it makes must be construed with caution. The denizens of the microworld with their "strangeness" and "charm" can hardly be said to be imaginable in the ordinary sense. At that level, we have lost many of the familiar bearings (such as individuality, sharp location, and measurement-independent properties) that allow us to anchor the reference of existence-claims in such macrotheories as geology or astrophysics. But imaginability must not be made the test for ontology. The realist claim is that the scientist is discovering the structures of the world; it is not required in addition that these structures be imaginable in the categories of the macroworld.

The form of the successful retroductive argument is the same at the micro- as at the macrolevel. If the success of the argument at the macrolevel is to be explained by postulating that something like the entities of the theory exist, the same ought to be true of arguments at the microlevel. Are there electrons? Yes, there are, just as there are stars and slowly moving geological plates bearing the continents of earth. What are electrons? Just what the theory of electrons says they are, no more, no less, always allowing for the likelihood that the theory is open to further refinement. If we cannot quite imagine what they are, this is due to the distance of the microworld from the world in which our imaginations were formed, not to the existential shortcomings of electrons (if I may so express the doubts of the antirealist).

A STRATEGY FOR SCIENTISTS?

Some of the critics of realism assume that defenders of realism are prescribing a strategy for scientists, a kind of regulative principle that will separate the good from the bad among proposed explanatory models. Since the critics believe this strategy to be defective, they have an additional argument against realism. In their view, nonrealist strategies as often as not work out. Indeed, two such episodes might be said to be foundational for modern science: Einstein's laying aside of ontological scruples in his rejection of classical space and time when formulating his general theory of relativity, and Heisenberg's restriction of matrix mechanics to observable quantities only.[11]

A contemporary example of a similarly non-realist strategy can be found among the proponents of S-matrix theory. Geoffrey Chew defends this approach against its rival, quantum field theory with its horde of theoretical entities, by claiming as an advantage that it has no "implication of physical meaning" and that its ability to dispense with an equation of motion allows it also to dispense with any sort of fundamental entities, such as particles or fields.[12] In some of his later essays, Heisenberg (the original proponent of the S-matrix formalism) dwelt on the choice facing quantum physicists of whether to opt for the Democritean approach, utilizing constituent entities, which has been canonical since the seventeenth century, or the Pythagorean approach, which relies on the resources of pure mathematics alone.[13] Heisenberg argued that the Pythagorean approach is now coming into its own, as the resources of the Democritean physical models are close to exhaustion at the quantum level.[14]

It is important to see why a realist could have supported Chew's effort and why the success of Heisenberg's early matrix mechanics must not be credited to antirealism. The realism/antirealism debate has to do with the assessment of the existential implications of successful theories *already in place.* It is not directed to strategies for *further* development, for deciding among alternative formalisms with respect to their likely future potential. A scientist who is persuaded of the truth of realism *might* very well decide that a fresh start is needed when he cannot find a coherent physical model around which to build a new theory. Positivism of this sort may well be called for in some situations, and the realist need not oppose it.

A realist might even decide that at some point the program of Heisenberg and Chew offers more promise, without repudiating his confidence in constructs that have been validated by earlier work. It is true, of course, that a realist will be less likely to turn in this direction than a non-realist would; the extended successes of the Democritean approach and the knowledge of physical structure it has made possible might weigh more heavily, as a sort of inductive argument, with the realist.

Nevertheless, there is no necessary connection; the defender of realism must not be saddled with a normative doctrine of the kind attributed here. One reason, perhaps, why this sort of confusion occurs is that Einstein's stand against Bohr is so often taken to be the paradigm of realism. And it did, indeed, involve a strongly normative doctrine in regard to the proper strategy for quantum physics. But Einstein's world view included, as I have shown, much more than realism; where it failed was not in its realistic component, but in the conservative constraints on future inquiry that Einstein felt the success of classical physics warranted.

As a footnote to this discussion, it may be worth emphasizing that the realist of whom I speak here is, in the first instance, a philosopher. The qualifier 'scientific' in front of 'realist' should not be allowed to mislead. It is used to distinguish the realism I am discussing from the many others that dot the history of philosophy. The realisms that philosophers in the past opposed to nominalism and to idealism are very different doctrines, and neither is connected, in any straightforward way at least, with the realism being referred to here. In the past, the realism I am speaking of has been most often contrasted with fictionalism or with 'instrumentalism'; but at this point the term is almost hopelessly equivocal.

'Scientific realism' is scientific because it proposes a thesis *in regard to* science. Though the case to be made for it may employ the inference-to-best-explanation technique also used in science, the doctrine itself is still a philosophic one. The scientist qua scientist is not called on to take a stand on it one way or another. Most scientists *do* have views on the issue, sometimes on the basis of much reflection but more often of a spontaneous kind. Indeed, it could be argued that worrying about whether or not their constructs approximate the real is more apt to hinder than to help their work as scientists!

SOURCES OF ANTIREALISM: HISTORY OF SCIENCE

The most obvious source of antirealism in recent decades is the new concern for the history of science on the part of philosophers of science. Thomas Kuhn's emphasis on the discontinuity that, according to him, characterizes the "revolutionary" transitions in the history of science also led him to a rejection of realism: "I can see [in the systems of Aristotle, Newton and Einstein] no coherent direction of ontological development."[15] Kuhn is willing to attribute a cumulative character to the low-level empirical laws of science. But he denies any cumulative character to theory: theories come and go, and many leave little of themselves behind.

Among the critics of realism, Larry Laudan is perhaps the one who sets most store in considerations drawn from the history of science. He displays an impressive list of once-respected theories that now have been discarded, and guesses that "for every highly successful theory in the past of science which we now believe to be a genuinely referring theory, one could find half a dozen once successful theories which we now regard as substantially non-referring."[16]

To meet this challenge adequately, it would be necessary to look closely at Laudan's list of discarded theories, and that would require an essay in its own right. But a few remarks are in order. The sort of theory on which the realist grounds his argument is one in which an increasingly finer specification of internal structure has been obtained over a long period, in which the theoretical entities function *essentially* in the argument and are not simply intuitive postulations of an "underlying reality," and in which the original metaphor has proved continuously fertile and capable of increasingly further extension. (More on this will follow.)

This excludes most of Laudan's examples right away. The crystalline spheres of ancient astronomy, the universal Deluge of catastrophist geology, theories of spontaneous generation—none of these would qualify. That is not to say that the entities or events they postulated were not firmly believed in by their proponents. But realism is not a blanket approval for all the entities postulated by long-supported theories of the past. Ethers and fluids are a special category, and one which Laudan stresses. I would argue that these were often, though not always, interpretive additions, that is, attempts to specify what "underlay" the equations of the scientist in a way which the equations (as we now see) did not really sanction. The optical ether, for example, in whose existence Maxwell had such confidence, was no more than a carrier for variations in the electromagnetic potentials. It seemed obvious that a vehicle of some sort *was* necessary; undulations cannot occur (as it was often pointed out) unless there is something to undulate! Yet nothing could be inferred about the carrier itself; it was an "I-know-not-what," precisely the sort of un-knowable "underlying reality" that the antirealist so rightly distrusts.

The theory of circular inertia and the effluvial theory of static electric-ity were first approximations, crude it is true, but effective in that the metaphors they suggested gradually were winnowed through, and some-thing of the original was retained. Phlogiston left its antiself, oxygen, behind. The view that the continents were static, which preceded the plate-tectonic model of contemporary geology, was not a theory; it was simply an assumption, one that is correct to a fairly high approximation. The early theories of the nucleus, which assumed it to be homogeneous, were simply idealizations; it was not known whether the nucleus was homogeneous or not, but a decision on that could be put off until first the notion of the nuclear atom itself could be fully explored. These are all examples given by Laudan. Clearly, they need more scrutiny than I have given them. But equally clearly, Laudan's examples may not be taken without further examination to count on the antirealist side. The value of this sort of reminder, however, is that it warns the realist that the onto-logical claim he makes is at best tentative, for surprising reversals *have* happened in the history of science. But the nonreversals (and a long list is easy to construct here also) still require some form of (philosophic) expla-nation, or so I shall argue.

SOURCES OF ANTIREALISM: PHILOSOPHY

According to the classic ideal of science as demonstration which domi-nated Western thought from Aristotle down to Descartes, hypothesis can be no more than a temporary device in science. Of course, one can find an abundance of retroductive reasoning in Aristotle's science as in Des-cartes', a tentative working back from observed effect to unobserved cause. But there was an elaborate attempt to ensure that *real* science, *sci-entia propter quid*, would not contain theoretical constructs of a hypo-thetical kind. And there was a tendency to treat these latter constructs as fictions, in particular the constructs of mathematical astronomy. Duhem has left us a chronicle of the antirealism with which the medieval phi-losophers regarded the epicycles and eccentrics of the Ptolemaic astron-omer.

EMPIRICISM

As the bar to hypothesis gradually came to be dropped in the seventeenth century, another source of opposition to theoretical constructs began to appear. The new empiricism was distrustful of unobserved entities, particularly those that were unobservable in principle. One finds this sort of skepticism already foreshadowed in some well-known chapters of Locke's *Essay Concerning Human Understanding*. Locke concluded there (Book IV) that a "science of bodies" may well be forever out of reach because there is no way to reason securely from the observed secondary qualities of things to the primary qualities of the minute parts on which those secondary qualities are supposed to depend. Hume went much further and restricted science to the patterning of sense impressions. He simply rejects the notion of cause according to which one could try to infer from these impressions to the unobserved entities causing them.

Kant tried to counter this challenge to the realistic understanding of Newtonian physics. He argued that entities such as the "magnetic matter pervading all bodies" need not be perceivable by the unaided senses in order to qualify as real.[17] He established a notion of cause sufficiently large to warrant causal inference from sense-knowledge to such unobservables as the "magnetic matter." Even though the transcendental deductions of the first *Critique* bear on the prerequisites of possible experience, 'experience' must be interpreted here as extending to all spatio-temporal entities that can be causally connected with the deliverances of our senses.[18]

Despite Kant's efforts, the skeptical empiricism of Hume has continued to find admirers. The logical positivists were attracted by it but were sufficiently impressed by the central role of theoretical constructs in science not to be quite so emphatic in their rejection of the reality of unobservable theoretical entities. The issue itself tended to be pushed aside and to be treated by them as undecidable; E. Nagel's *The Structure of Science* gives classical expression to this view. This sort of agnosticism alternated with a more definitely skeptical view in logical positivist writings. If one takes empiricism as a starting point, it is tempting to push it (as Hume did) to yield the demand not just that every claim about the world must ultimately rest on sense experience but that every admissible entity must be directly certifiable by sense experience.

This is the position taken by Bas van Fraassen. His antirealism is restricted to those theoretical entities that are in principle unobservable. He has no objection to allowing the reality of such theoretical entities as stars (interpreted as large glowing masses of gas) because these are, in his view, observable in principle since we could approach them by spaceship, for example. It is part of what he calls the "empirical adequacy" of a stellar theory that it should predict what we would observe should we come to a star. This criterion, which he makes the single aim of science, is sufficiently broad, therefore, to allow reality-claims for any theoretical entity that, though at present unobserved, is at least in principle directly observable by us. His antirealism has more than a tinge of old-fashioned nominalism about it, the rejection of what he calls an "inflationary meta-physics" of redundant entities.[19] Since neither of the two main arguments he lists for realism, inference to the best explanation and the common cause argument, are (in his view) logically compelling, this is taken to justify his application of Occam's razor.

One immediate difficulty with this position is, of course, the distinc-tion drawn between the observable and the unobservable. Since entities on one side of the line are ontologically respectable and those on the other are not, it is altogether crucial that there be some way not only to draw the distinction but also to confer on it the significance that van Fraassen attributes to it. In one of the classic papers in defense of scien-tific realism, Grover Maxwell argued in 1962 that there is a continuum in the spectrum of observation from ordinary unaided seeing down to the operation of a high-power microscope.[20] Van Fraassen concedes that the distinction is not a sharp one, that 'observe' is a vague predicate, but insists that it is sufficient if the ends of the spectrum be clearly distinct, that is, that there be at least some clear cases of supposed interaction with theoretical entities which would not count as "observing."[21] He takes the operation of a cloud chamber, with its ionized tracks allegedly indicating the presence of charged entities such as electrons, to be a case where "observe" clearly ought not be used. One must not say, on noting such a track: I observed an electron.

To lay as much weight as this on the contingencies of the human sense organs is obviously problematic, as van Fraassen recognizes. There are organisms with sense-organs very different from ours that can perceive phenomena such as ultraviolet light or the direction of optical polariza-tion. Why could there not, in principle, be organisms much smaller than we, able to perceive microentities that for us are theoretical and able also to communicate with us? Is not the notion 'observable in principle' hope-lessly vague in the face of this sort of objection? How can it be used to draw a usable distinction between theoretical entities that do have onto-logical status and those that do not? Van Fraassen's response is cautious:

It is, on the face of it, not irrational to commit oneself only to a search for theo-ries that are empirically adequate, ones whose models fit the observable phenom-ena, while recognizing that what counts as an observable phenomenon is a func-tion of what the epistemic community is (that *observable* is *observable-to-us*).[22]

So 'observable' means here "observable in principle by us with the sense organs we presently have." But once again, why would 'unobserv-able' in this sense be allowed the implications for epistemology and ontology that van Fraassen wants to attach to it?[23] The question is not whether the aim of science ought to be broadened to include the search for unobservable but real entities, though something could be said in favor of such a proposal. It is sufficient for the purposes of the realist to ask whether theories that are in van Fraassen's sense empirically adequate can also be shown under certain circumstances to have likely ontological implications.

Van Fraassen allows that the moons of Jupiter can be observed through a telescope; this counts as observation proper "since astronauts will no doubt be able to see them as well from close up."[24] But one cannot be said to "observe" by means of a high-power microscope (he alleges) because no such direct alternative is available to us in this case. What matters here is not so much the way the instrument works, the precise physical or theoretical principles involved. It is whether there is also, in principle, a direct unmediated alternative mode of observation available to us. The entity need not be observable *in practice.* The iron core that geologists tell us lies at the center of the earth is certainly not observable in practice; it is a theoretical entity since its existence is known only through a successful theory, but it may nonetheless be regarded as real, van Fraassen would say, because *in principle* we could go down there and check it out.

The quality of the evidence for this geological entity might, however, seem no better than that available for the chromosome viewed by microscope. Van Fraassen rests his case on an analysis of the aims of science, in an abstract sense of the term 'aim', on the "epistemic attitude" (as he calls it) proper to science as an activity. And he thinks that reality-claims in the case of the chromosomes, but not the iron core, lie outside the permissible aims of science. Is there any way to make this distinction more plausible?

REFERENCE

Some theoretical entities (such as the iron core or the star) are of a kind that is relatively familiar from other contexts. We do not need a theory to tell us that iron exists or how it may be distinguished. But electrons are what quantum theory says they are, and our only warrant for knowing that they exist is the success of that theory. So there is a special class of theoretical entities whose *entire* warrant lies in the theory built around them. They correspond more or less to the unobservables of van Fraassen.

What makes them vulnerable is that the theory postulating them may itself change or even be dropped. This is where the problems of meaning change and of theory replacement so much discussed in recent philosophy of science become relevant. The antirealist might object to a reality-claim for electrons or genes not so much because they are unobservable but because the reference of the term 'electron' may shift as theory changes. To counter this objection, it sounds as though the realist will have to provide a theory of reference that is able to secure a constancy of reference in regard to such theoretical terms. R. Rorty puts it this way:

The need to pick out objects without the help of definitions, essences, and meanings of terms, produced (philosophers thought) a need for a "theory of reference" which would not employ the Fregean machinery which Quine had rendered dubious. This call for a theory of reference became assimilated to the demand for a "realistic" philosophy of science which would reinstate the pre-Kuhnian and pre-Feyerabendian notion that scientific inquiry made progress by finding out more and more about the same objects.[25]

Rorty is, of course, skeptical of theories of reference generally, and derides the idea that the problems of realism could be handled by such a theory. He chides Putnam, in particular, for leading philosophers to believe that they could be. Recall the celebrated realist's nightmare conjured up by Putnam:

What if all the theoretical entities postulated by one generation (molecules, genes, etc. as well as electrons) invariably "don't exist" from the standpoint of later science? . . . One reason this is a serious worry is that eventually the following meta-induction becomes compelling: just as no term used in the science of more than 50 (or whatever) years ago referred, so it will turn out that no term used now (except maybe observation-terms if there are such) refers.[26]

This is the "disastrous meta-induction" which at that time Putnam felt had to be blocked at all costs. But, of course, if the theoretical entities of one generation really did *not* have any existential claim on the next, realism simply would be false. It is, in part at least, because the history of science testifies to a substantial continuity in theoretical structures that we are led to the doctrine of scientific realism at all. Were the history of science *not* to do so, then we would have no logical or metaphysical grounds for believing in scientific realism in the first place. But this is to get ahead of the story. I introduced the issue of reference here not to argue its relevance one way or the other, but to note that one form of antirealism can be directed against the subset of theoretical entities which derive their definition entirely from a particular theory.

One way for a realist to evade objections of this kind is to focus on the manner in which theoretical entities can be causally connected with our measurement apparatus. An electron may be defined as the entity that is causally responsible for, among other things, certain kinds of cloud tracks. A small number of parameters, such as mass and charge, can be associated with it. Such an entity will be said to exist, that is, not to be an artifact of the apparatus, if a number of convergent sorts of causal lines lead to it. There would still have to be a theory of some sort to enable the causal tracking to be carried out. But the reason to affirm the entity's existence lies not in the success of the theory in which it plays an explanatory role, but in the operation of traceable causal lines. Ian Hacking urges that this defense of realism, which relies on experiential interactions, avoids the problems of meaning-change that beset arguments based on inference to the best explanation.[27]

TRUTH AS CORRESPONDENCE

The most energetic criticisms of realism, of late, have been coming from those who see it as the embodiment of an old-fashioned, and now (in their view) thoroughly discredited, attachment to the notion of truth as some sort of "correspondence" with an "external world." These criticisms take quite different forms, and it is impossible to do them justice in a short space. The rejected doctrine is one that would hold that even in the ideal limit, the best scientific theory, one that has all the proper methodological virtues, could be false. This embodies what the critics have come to call the "God's eye view," the view that there may be more to the world than our language and our sciences can, even in principle, express. They concede that the doctrine has been a persuasive one ("it is impossible to find a philosopher before Kant who was *not* a metaphysical realist");[28] its denial seems, indeed, shockingly anthropomorphic. But they are in agreement that no philosophic sense can be made of the central metaphor of correspondence: "To single out a correspondence between two domains, one needs some independent access to both domains."[29] And, of course, an independent "access to the noumenal objects" is impossible.

The two main protagonists of this view are, perhaps, Rorty and Putnam. Rorty is the more emphatic of the two. He defends a form of pragmatism that discounts the traditional preoccupations of the philosopher with such Platonic notions as truth and goodness. He sees the Greek attempt to separate *doxa* and *epistēmē* as misguided; he equally refuses the modern trap of trying to analyze the meaning of 'true', because it would involve an "impossible attempt to step outside our skins."[30] The pragmatist

drops the notion of truth as correspondence with reality altogether, and says that modern science does not enable us to cope because it corresponds, it just plain enables us to cope. His argument for the view is that several hundred years of effort have failed to make interesting sense of the notion of "correspondence," either of thoughts to things or of words to things.[31]

Does Rorty deny scientific realism, that is, the view that the long-term success of a scientific theory gives us a warrant to believe that the entities it postulates do exist? It is not clear. What is clear, first, is that he rejects any kind of argument for scientific realism that would explain the success of a theory in terms of a correspondence with the real. And second, he denies that scientific claims have a privileged status, that the scientists' table (in Eddington's famous story) is the only real table. Science, he retorts, is just "one genre of literature," a way "to cope with various bits of the universe," just as ethics helps us cope with other bits.[32]

Putnam, in contrast, is willing to ask the traditional philosophic questions. His patron is Kant rather than James.[33] 'Truth' he defines as "an idealization of rational acceptability."[34] He has more specific objections to urge against the offending metaphysical version of realism than does Rorty, whose argument amounts to claiming that it has failed to make "interesting sense."[35] Does he link this rejection with a rejection of scientific realism? Certainly not in his *Meaning and the Moral Sciences* (1978), where he defends scientific realism by urging that it permits the best explanation of the success of science. It is somewhat more difficult to be sure where his allegiances lie in his more recent pieces; his earlier enthusiasm for scientific realism seems, at the least, to be waning.[36] He attacks materialism with its assumption of mind-independent things,[37] as well as reductionism.

We are too realistic about physics . . . [because] we see physics (or some hypothetical future physics) as the One True Theory, and not simply as a rationally acceptable description suited for certain problems and purposes.[38]

This does not sound like scientific realism. Be this as it may, however, it seems clear that scientific realism is not the main target in this debate. The target is a set of metaphysical views, views (it is true) that scientific realists have in the past usually taken for granted. I suspect that Rorty would allow that genes exist and that dinosaurs once roamed the earth, as long as these claims are not given a status that is denied to more mundane statements about chairs and goldfish. But can we allow him this position so easily?

Recall that the original motivation for the doctrine of scientific realism was not a perverse philosopher's desire to inquire into the unknowable or to show that only the scientist's entities are "really real." It was a response to the challenges of fictionalism and instrumentalism, which over and over again in the history of science asserted that the entities of the scientist are fictional, that they do not exist in the everyday sense in which chairs and goldfish do. Now, how does Rorty respond to this? Has he an argument to offer? If he has, it would be an argument for scientific realism. It would also (as far as I can see) be a return to philosophy in the "old style" that he thinks we ought to have outgrown.

My own inclinations are to defend a form of metaphysical realism, though not necessarily under all the diverse specifications Putnam offers of it.[39] But that is not to the point here. What is to the point is that scientific realism is not immediately undermined by the rejection of metaphysical realism, though the character of the claim scientific realism makes obviously depends on whether or not it is joined to a concept of truth in which the embattled notion of "correspondence" plays a part. Further, the type of argument most often alleged in its support *does* use the language of correspondence: it is the approximate correspondence between the physical structure of the world and postulated theoretical entities that is held to explain why a theory succeeds as well as it does.[40] Readers will have to decide for themselves whether my argument below does "make interesting sense" or not.

VARIETIES OF ANTIREALISM

It may be worthwhile at this point, looking back at the territory we have traversed, to draw two rough distinctions between types of antirealism. *General antirealism* denies ontological status to theoretical entities of science generally, while *limited antirealism* denies it only to certain classes of theoretical entities, such as those that are said to be unobservable in principle. Thus, the arguments of Laudan, based as they are on a supposedly general review of the history of scientific theories, would lead him to a *general* form of antirealism, one that would exclude existence status to *any* theoretical entity whose existence is warranted only by the success of the theory in which it occurs. In contrast, van Fraassen is claiming, as I have shown, only a *limited* form of antirealism.

Second, we might distinguish between *strong antirealism*, which denies any kind of ontological status to all (or part) of the theoretical entities of science, and *weak antirealism* which allows theoretical entities existence of an everyday "chairs and goldfish" kind,[41] but insists that there is some further sense of "really really there," which realists purportedly have in mind, that is to be rejected. Classical instrumentalism would be of the former kind (strong antirealists), whereas many of the more recent critics of scientific realism appear to fall in the latter category (weak antirealists). These (weak antirealist) critics are often, as I have shown, hard to place. They reject any attempt to justify scientific realism as involving dubious metaphysics, but appear to accept a weak (realist) claim of the "everyday" kind without any form of supporting argument.[42] Their rhetoric is antirealist in tone, but their position often seems compatible with the most basic claim of scientific realism, namely that there is reason to believe that the theoretical terms of successful theories refer. This gives the weak antirealists' position a puzzling sort of undeclared status where they appear to have the best of both worlds. I am inclined to think that their effort to have it both ways must in the end fail.

THE CONVERGENCES OF STRUCTURAL EXPLANATION

The basic claim made by scientific realism, once again, is that the long-term success of a scientific theory gives reason to believe that something like the entities and structure postulated by the theory actually exists. There are four important qualifications built into this: (1) the theory must be successful over a significant period of time; (2) the explanatory success of the theory gives some reason, though not a conclusive warrant, to believe it; (3) what is believed is that the theoretical structures are *something like* the structure of the real world; (4) no claim is made for a special, more basic, privileged, form of existence for the postulated entities.[43] These qualifications: "significant period," "some reason," "something like," sound very vague, of course, and vagueness is a challenge to the philosopher. Can they not be made more precise? I am not sure that they can; efforts to strengthen the thesis of scientific realism have, as I have shown, left it open to easy refutation.

The case for scientific realism can be made in a variety of ways. Maxwell, Salmon, Newton-Smith, Boyd, Putnam, and others have argued it in well-known essays. I am not going to comment on their arguments here since my aim is to outline what I think to be the best case for scientific realism. My argument will, of course, bear many resemblances to theirs. What may be the most distinctive feature of my argument is my stress on structural types of explanation, and on the role played by the criterion of fertility in such explanations.

Stage one of the argument will be directed especially against general antirealism. I want to argue that in many parts of natural science there has been, over the last two centuries, a progressive discovery of *structure*. Scientists construct theories which explain the observed features of the physical world by postulating models of the hidden structure of the entities being studied. This structure is taken to account causally for the observable phenomena, and the theoretical model provides an approximation of the phenomena from which the explanatory power of the model derives. This is the standard account of structural explanation, the type of explanation that first began to show its promise in the eighteenth and early nineteenth centuries in such sciences as geology and chemistry.[44]

I want to consider some of the areas where the growth in our knowledge of structure has been relatively steady. Let me begin with geology, a good place for a realist to begin. The visible strata and their fossil contents came to be interpreted as the evidence for an immense stretch of time past in which various processes such as sedimentation and volcanic activity occurred. There was a lively debate about the mechanisms of mountain building and the like, but gradually a more secure knowledge of the past aeons built up. The Carboniferous period succeeded the Devonian and was, in turn, succeeded by the Permian. The length of the periods, the climatic changes, and the dominant life forms were gradually established with increasing accuracy. It should be stressed that a geological period, such as the Devonian, is a theoretical entity. Further, it is, in principle, inaccessible to our direct observation. Yet our theories have allowed us to set up certain temporal boundaries, in this case (the Devonian period) roughly 400 to 350 million years ago, when the dominant life form on earth was fish and a number of important developments in the vertebrate line occurred.

The long-vanished species of the Devonian are theoretical entities about which we have come to know more and more in a relatively steady way. Of course, there have been controversies, particularly over the sudden extinction of life forms such as occurred at the end of the Cretaceous period and over the precise evolutionary relationships among given species. But the very considerable theory changes that have occurred since Hutton's day do not alter the fact that the growth in our knowledge of the sorts of life forms that inhabited the earth aeons ago has been pretty cumulative. The realist would say that the success of this synthesis of geological, physical, and biological theories gives us good reason to believe that species of these kinds did exist at the times and in the conditions proposed. Most antirealists (I suspect) would agree. But if they do, they must concede that this mode of retroductive argument can warrant, at least in some circumstances, a realist implication.

Geologists have also come to know (in the scientists' sense of the term 'know') a good deal about the interior of the earth. There is a discontinuity between the material of the crust and the much denser mantle, the "Moho" as it is called after its Yugoslavian discoverer, about 5 kilometers under the ocean bed and much deeper, around 30 or 40 kilometers, under the continents. There is a further discontinuity between the solid mantle and the molten core at a depth of 2,900 kilometers. All this is inferred from the characteristics of seismic waves at the surface. Does this structural model of the earth simply serve as a device to enable the scientist to predict the seismic findings more accurately, or does it enable an additional ontological claim to be made about the actual hidden structures of earth? The realist would argue that the explanatory power of the geologist's hypothesis, its steadily improving accuracy, gives good ground to suppose that something can be inferred about real structures that lie far beneath us.

An elegant example of a quite different sort would come from cell biology. Here, the techniques of microscopy have interwoven with the theories of genetics to produce an ever more detailed picture of what goes on inside the cell. The chromosome first appeared under a microscope; only gradually was the gene, the theoretical unit of hereditary transmission, linked to it. Later the gene came to be associated with a particular locus on the chromosome. The unraveling by Crick and Watson of the biochemical structure of the chromosome made it possible to define the structure of the gene in a relatively simple way and has allowed at least the beginnings of an understanding of how the gene operates to direct the growth of the organism. In his book, *The Matter of Life*, Michael Simon has traced this story in some detail, and has argued that its progressive character can best be understood in terms of a realist philosophy of science.[45]

One further example of this sort of progression can be found in chemistry. The complex molecules of both inorganic and organic chemistry have been more accurately charted over the past century. The atomic constituents and the spatial relations among them can be specified on the basis both of measurement, using X-ray diffraction patterns, for example, and on the basis of a theory that specifies where each kind of atom *ought* to fit. Indeed, this knowledge has enabled a computer program to be designed that can "invent" molecules, can suggest that certain configurations would yield a new type of complex molecule and can even predict what some of the molecule's properties are likely to be.

To give a realist construal to the molecular models of the chemist is not to imply that the nature of the constituent atoms and of the bonding between them is exhaustively known. It is only to suppose that the elements and spatial relationships of the model disclose, in a partial and tentative way, real structures within complex molecules. These structures are coming to be more exactly charted, using a variety of techniques both experimental and theoretical. The coherence of the outcome of these widely different techniques, and the reliability of the chemist's intuitions as he decides which atom must fit a particular spot in the lattice, are most easily understood in terms of the realist thesis.

These examples may serve to make two points. The first is that the discontinuous replacement account of the history of theories favored by antirealists is seen to be one-sided. If one focuses on global explanatory theories, particularly in mechanics, it can come to seem that theoretical entities are modified beyond recognition as theories change. Dirac's electron has little in common with the original Thomson electron; Einstein's concept of time is a long way from Newton's, and so on. These conventional examples of conceptual change could themselves be scrutinized to see whether they will bear the weight the antirealist gives them. But it may be more effective to turn from explanatory elements such as electrons to explanatory structures such as those of the organic chemist, and note, as a historical fact, the high degree of continuity in the relevant history.

Second, one could note the sort of confidence that scientists have in structural explanations of this sort. It is not merely a confidence in the empirical adequacy of the predictions these models enable them to make. It is a confidence in the model itself as an analysis of complex real structure. Look at any textbook of polymer chemistry to verify this. Of course, the chemists could be wrong to build this sort of realist expectation into their work, but the arguments of philosophers are not likely to convince them of it.

A third consequence one might draw from the history of the structural sciences is that there is a single form of retroductive inference involved throughout. As C. S. Peirce stressed in his discussion of retroduction, it is the degree of success of the retroductive hypothesis that warrants the degree of its acceptance as truth. The point is a simple one, and indeed is already implicit in Aristotle's *Posterior Analytics*. Aristotle indicates that what certifies as *demonstrative* a piece of reasoning about the relation between the nearness of planets and the fact that they do not twinkle, is the degree to which the reasoning *explains*. This connection between the explanatory and the epistemic character of scientific reasoning is constantly stressed in Renaissance and early modern discussions of hypothetical reasoning.[46]

What the history of recent science has taught us is not that retroductive inference yields a plausible knowledge of causes. We already knew this on *logical* grounds. What we have learned is that retroductive inference *works* in the world we have and with the senses we have for investigating that world. This is a contingent fact, as far as I can see. This is why realism as I have defined it is in part an empirical thesis. There could well be a universe in which observable regularities would *not* be explainable in terms of hidden structures, that is, a world in which retroduction would not work. Indeed, until the eighteenth century, there was no strong empirical case to be made against that being *our* universe. Scientific realism is not a logical doctrine about the implications of successful retroductive inference. Nor is it a metaphysical claim about how any world *must* be. It has both logical and metaphysical components. It is a quite limited claim that purports to explain why certain ways of proceeding in science have worked out as well as they (contingently) have.

That they have worked out well in such structural sciences as geology, astrophysics, and molecular biology, is apparent. And the presumption in these sciences is that the model-structures provide an increasingly accurate insight into the real structures that are causally responsible for the phenomena being explained. This may be thought to give a reliable presumption in favor of the realist implications of retroductive inference in natural science generally. But one has to be wary here. Much depends on the sort of theoretical entity one is dealing with; I have already noted, for instance, some of the perplexities posed by quantum-mechanical entities. Much depends too on how *well* the theoretical entity has served to explain: How important a part of the theory has it been? Has it been a sort of optional extra feature like the solid spheres of Ptolemaic astronomy? Or has it guided research in the way the Bohr model of the hydrogen atom did? What kind of fertility has the theoretical entity shown?

FERTILITY AND METAPHOR

Kuhn lists five values that scientists look for when evaluating a scientific theory: predictive accuracy, consistency, breadth of scope, simplicity, fertility.[47] It is the last of these that bears most directly on the problem of realism. Fertility is usually equated with the ability to make novel predictions. A good theory is expected to predict novel phenomena, that is, phenomena that were not part of the set to be explained. The further in kind these novel phenomena are from the original set, and thus the more unexpected they are, the better the model is said to be. The display of this sort of fertility reduces the likelihood of the theory's being an ad hoc one, one invented just for the original occasion but with no further scope to it.

There has been much debate about the significance of this notion of ad hoc. Clearly, it will appeal to the realist and will seem arbitrary to the antirealist. The realist takes an ad hoc hypothesis not to be a genuine theory, that is, not to give any insight into real structure and therefore to have no ground for further extension. The fact that it accounts for the original data is accidental and testifies to the ingenuity of the inventor rather than to any deeper fit. When the theory is first proposed, it is often difficult to tell whether or not it is ad hoc on the basis of the other criteria of theory appraisal. This is why fertility is so important a criterion from the realist standpoint.

The antirealist will insist that the novel facts predicted by the theory simply increase its scope and thus make it more acceptable. They will say that there is no significance to the time order in which predictions are made; if they are successful, they count as evidence whether or not they pertain to the data originally to be explained. A straightforward application of Bayes's theorem shows this, assuming of course the antirealist standpoint. Yet scientists seem to set a lot of store in the notion of ad hoc. Are scientific intuitions sufficiently captured by a translation into antirealist language? Is an ad hoc hypothesis one that just happens not to be further generalizable, or is it one that does not give sufficient insight into real structure to permit any further extension?

Rather than debate this already much-debated issue further, let me turn to a second aspect of fertility which is less often noted but which may be more significant for our problem.[48] The first aspect of fertility, novelty, had to do with what could logically be inferred from the theory, its logical resources, one might put it. But a good model has more resources than these. If an anomaly is encountered or if the theory is unable to predict one way or the other in a domain where it seems it *should* be able to do so, the model itself may serve to suggest possible modifications or extensions. These are *suggested*, not implied. Therefore, a creative move on the part of the scientist is required.

In this case, the model functions somewhat as a metaphor does in language. The poet uses a metaphor not just as decoration but as a means of expressing a complex thought. A good metaphor has its own sort of precision, as any poet will tell you. It can lead the mind in ways that literal language cannot. The poet who is developing a metaphor is led by suggestion, not by implication; the reader of the poem queries the metaphor and searches among its many resonances for the ones that seem best to bear insight. The simplistic "man is a wolf" examples of metaphor have misled philosophers into supposing that what is going on in metaphor is a comparison between two already partly understood things. The only challenge then would be to decide in what respects the analogy holds. In the more complex metaphors of modern poetry, something much more interesting is happening. The metaphor is helping to illuminate something that is not well understood in advance, perhaps, some aspect of human life that we find genuinely puzzling or frightening or mysterious. The manner in which such metaphors work is by tentative suggestion. The minds of poet and reader alike are actively engaged in creating. Obviously, much more would need be said about this, but it would lead me too far afield at this point.[49]

The good model has something of this metaphoric power.[50] Let me recall another one here, from geology once again. It had long been known that the west coast of Africa and the east coast of South America show striking similarities in terms of strata and their fossil contents. In 1915, Alfred Wegener put forward a hypothesis to explain these and other similarities, such as those between the major systems of folds in Europe and North America. The continental drift notion that he developed in *The Origins of Continents and Oceans* was not at first accepted, although it admittedly did explain a great deal. There were too many anomalies: How could the continents cut through the ocean floor, for example, since the material of the ocean floor is considerably harder than that of the continents? In the 1960s, new evidence of seafloor spreading led H. Hess and others to a modification of the original model. The moving elements are not the continents but rather vast plates on which the continents as well as the seafloor are carried. And so the continental drift hypothesis developed into the plate tectonic model.

The story has been developed so ably from the methodological standpoint by Rachel Laudan[51] and Henry Frankel[52] that I can be very brief, and simply refer you to their writings. The original theoretical entity, a floating continent, did not logically entail the plates of the new model. But in the context of anomalies and new evidence, it did *suggest* them. And these plates in turn suggested new modifications. What happens when the plates pull apart are seafloor rifts, with quite specific properties. The upwelling lava will have magnetic directional properties that will depend on its orientation relative to the earth's magnetic field at the time. This allows the lava to be dated, and the gradual pulling apart of the plates to be charted. It was the discovery of such dated strips paralleling the midocean rifts that proved decisive in swinging geologists over to the new model in the mid-1960s. What happens when the plates collide? One is carried down under (subduction); the other may be upthrust to form a mountain ridge. One can see here how the original metaphor is gradually extended and made more specific.

In a recent critical discussion of my views on fertility and metaphor,[53] Michael Bradie has urged as a weakness of my argument that one needs to give a sufficiently precise account of metaphor to allow one to understand what would count as a metaphorical extension, so as to know when two theory stages can be identified as different stages of the same theory. My response is simple and, perhaps, simplistic. If the original model (say, continental drift) suggested the later modification as a plausible way of meeting the known anomalies and of incorporating the new evidence, then I would call this a metaphorical extension. Are continental drift and the plate tectonic model two stages of the same theory or two different theories? It all depends on how 'theory' is defined and how sharply theories are individuated. I do not see that very much hangs on this decision, one way or the other.

The important thing to note is that there *are* structural continuities from one stage to the next, even though there are also important structural modifications. What provides the continuity is the underlying metaphor of moving continents that had been in contact a long time ago and had very gradually developed over the course of time. One feature of the original theory, that the continents are the units, is eventually dropped; other features, such as what happens when the floating plates collide, are thought through and made specific in ways that allow a whole mass of new data to fall into place.

How does all this bear on the argument for realism? The answer should be obvious. This kind of fertility is a persistent feature of structural explanations in the natural sciences over the last three centuries and especially during the last century. How can it best be understood? It appears to be a contingent feature of the history of science. There seems to be no a priori reason why it *had* to work out that way, as I have already shown. What best explains it is the supposition that the model approximates sufficiently well the structures of the world that are causally responsible for the phenomena to be explained to make it profitable for the scientist to take the model's metaphoric extensions seriously. It is because there is something like a floating plate under our feet that it is proper to ask: What happens when plates collide, and what mechanisms would suffice to keep them in motion? These questions do not arise from the original theory if it is taken as no more than a formalism able to give a reasonably accurate predictive account of the data then at hand. If the continental drift hypothesis had no implications for what is really going on beneath us, for the hidden structures responsible for the phenomena of the earth's surface, then the subsequent history of that hypothesis would be unintelligible. The antirealist cannot, it seems to me, make sense of such sequences, which are pretty numerous in the recent history of all the natural sciences, basic mechanics, as always, constituting a special case.

One further point is worth stressing in regard to our geological story. Some theoretical features of the model, such as the midocean rifts, could be checked directly and their existence observationally shown. Here, as so often in science, theoretical entities previously unobserved, or in some cases even thought to be unobservable, are in fact observed and the expectations of theory are borne out, to no one's surprise. The separation between observable and unobservable postulated by many antirealists in regard to ontological status does not seem to stand up. The same mode of argument is used in each case; it is not clear why in one case expectations of real existence are accorded to the theoretical entity whereas in other cases, logically similar in explanatory character, these expectations are denied. The ontological inference, let me insist again, must be far more hesitant in some cases than in others. There is no question of according the same ontological status to *all* theoretical entities by virtue of a similar degree of fertility evinced over a significant period of time. Nonetheless, such fertility finds its best explanation in a broadly realist account of science.

Does this form of argument commit the realist to holding that every regularity in the world must be explained in terms of ontological structure? This turns out to be van Fraassen's main line of attack against realism. He takes it that the realist is committed to finding hidden variables in quantum mechanics. Since the odds against this are now quite high, and since, in any event, this would commit the realist to one possible world where the other looks just as possible, van Fraassen takes this to refute realism. But as I have shown, realism is not a regulative principle, and it does not lay down a strategy for scientists. Realism would not be refuted if the decay of individual radioactive atoms turns out to be genuinely undetermined. It does not look to the future; much more modestly, realism looks to quite specific past historical sequences and asks what best explains them. Realism does not look at *all* science, nor at all future science, just at a good deal of past science which (let me say it again) might not have worked out to support realism the way it did. The realist seeks an explanation for the regularities he finds in science, just as the scientist seeks an explanation for regularities he finds in the world. But if in particular cases he cannot find an explanation or cannot even show that there is no explanation, this in no sense shows that his original aim has somehow been discredited.

Thus, what van Fraassen describes as the "nominalist response" of the antirealist must in the end be rejected. He characterizes it in this way:

That the observable phenomena exhibit these regularities, because of which they fit the theory, is merely a brute fact, and may or may not have an explanation in terms of unobservable facts 'behind the phenomena'—it really does not matter to the goodness of the theory, nor to our understanding of the world.[54]

I hope I have shown that the nominalist resolve to leave such regularities as the extraordinary fertility of our scientific theories at the level of brute fact is unphilosophical. Furthermore, I hope I have shown that it makes a very great deal of difference to the explanatory power or goodness of a theory whether it can call on effective metaphors of hidden structure. And I doubt whether it is really necessary to prove that such metaphors are important to our understanding of the world and of the role of science in achieving such understanding.

EPILOGUE

Finally, I return to the weighty issues of reference and truth which are so dear to the heart of the philosopher. Clearly, my views on metaphor would lead me to reject the premise on which so much of the recent debate on realism has been based. Van Fraassen puts it thus:

Science aims to give us, in its theories, a literally true story of what the world is like; and acceptance of a scientific theory involves the belief that it is true. This is the correct statement of scientific realism.[55]

I do not think that acceptance of a scientific theory involves the belief that it is true. Science aims at fruitful metaphor and at ever more detailed structure. To suppose that a theory is literally true would imply, among other things, that no further anomaly could, in principle, arise from any quarter in regard to it. At best, it is hard to see this as anything more than an idealized "horizon-claim," which would be quite misleading if applied to the actual work of the scientist. The point is that the resources of metaphor are essential to the work of science and that the construction and retention of metaphor must be seen as part of the aim of science.

Scientists in general accept the quantum theory of radiation. Do they believe it to be true? Scientists are very uncomfortable at this use of the word 'true', because it suggests that the theory is definitive in its formulation. As has often been pointed out, the notion of *acceptance* is very complex, indeed ambiguous. It is basically a pragmatic notion: one accepts an explanation as the best one available; one accepts a theory as a good basis for further research, and so forth. In no case would it be correct to say that acceptance of a theory entails belief in its truth.

The realist would not use the term 'true' to describe a good theory. He would suppose that the structures of the theory give some insight into the structures of the world. But he could not, in general, say how good the insight is. He has no independent access to the world, as the antirealist constantly reminds him. His assurance that there is a fit, however rough, between the structures of the theory and the structures of the world comes not from a comparison between them but from the sort of argument I sketched above, which concludes that only this sort of reasoning would explain certain contingent features of the history of recent science. The term 'approximate truth', which has sometimes been used in this debate, is risky because it immediately invites questions such as: *how approximate?*, and how is the degree of approximation to be measured? If I am right in my presentation of realism, these questions are unanswerable because they are inappropriate.

The language of theoretical explanation is of a quite special sort. It is open-ended and ever capable of further development. It is metaphoric in the sense in which the poetry of the symbolists is metaphoric, not because it uses explicit analogy or because it is imprecise, but because it has resources of suggestion that are the most immediate testimony of its ontological worth. Thus, the M. Dummett-Putnam claim that a realist is committed to holding with respect to any given theory, that the sentences of the theory are either true or false,[56] quite misses the mark where scientific realism is concerned. Indeed, I am tempted to say (though this would be a bit too strong) that if they are literally true or false, they are not of much use as the basis for a research program.

Ought the realist be apologetic, as his pragmatist critic thinks he should be, about such vague-sounding formulations as these: that a good model gives an insight into real structure and that the long-term success of a theory, in most cases, gives reason to believe that something like the theoretical entities of that theory actually exist? I do not think so. The temptation to try for a sharper formulation must be resisted by the realist, since it would almost certainly compromise the sources from which his case derives its basic strength. And the antirealist must beware of the opposite temptation to suppose that whatever cannot be said in a semantically definitive way is not worth saying.

NOTES

The first version of this essay was delivered as an invited paper at the Western Division meeting of the American Philosophical Association in April 1981. I am indebted to Larry Laudan for his incisive commentary on that occasion, and to the numerous discussions we have had on this topic.

1. It was the confidence that, as a student of physics, I had developed in this belief that led me, in my first published paper in philosophy, to formulate a defense of scientific realism against the instrumentalism prevalent at the time among philosophers of science. (See "Realism in Modern Cosmology," *Proceedings American Catholic Philosophical Association* 29 [1955]: 137-150.) Much has changed in philosophy of science since that time; a different sort of defense is (as we shall see) now called for.

2. This is the theme of C. G. Hempel's classic essay, "The Theoretician's Dilemma," *Minnesota Studies in the Philosophy of Science* 3 (1958): 37-98.

3. For the details of this story, see E. McMullin, *Newton on Matter and Activity* (Notre Dame: University of Notre Dame Press, 1978), especially chap. 4: "How is Matter Moved?"

4. In a recent critique of "metaphysical realism," Hilary Putnam has Newton defending the view that particles act at a distance across empty space. *Reason, Truth and History* (Cambridge: Cambridge University Press, 1981), 73. Though the *Principia* has often been made to yield that claim, this view is, in fact, the one alternative that Newton at all times steadfastly rejected.

5. Newton's other suggestion, briefly explored in the 1690s, that forces might be nothing other than the manifestations of God's direct involvement in the governance of the universe, *could*, however, be properly described as 'metaphysical'; this is not, of course, to say that it was illegitimate.

6. H. Putnam, "Why There Isn't a Ready-Made World," *Synthese* 51 (1982): 141-168; see 163. Also available in volume 3 of Putnam's Philosophical Papers Series, *Realism and Reason* (Cambridge: Cambridge University Press, 1983).

7. According to Putnam, Newton, though no positivist, "strongly rejected the idea that his theory of universal gravitation could or should be read as a description of metaphysically ultimate fact. '*Hypotheses non fingo*' was a rejection of metaphysical hypotheses, not of scientific ones" (*Reason, Truth and History*, 163). This supposed rejection of metaphysics would, however, place Newton much closer to positivism than he really was. In the *Principia*, Newton shows himself well aware that different interpretations (he calls them "physical," not "metaphysical") can be given of attraction, and he tries to deflect anticipated criticism of this ambiguity by intimating that one can prescind such interpretation by remaining at the "mathematical" level. But he knew perfectly well that he could not *remain* at this level and still claim to have "explained" the planetary motions. In his own later writing, much of it unpublished in his lifetime, he constantly tried out different hypotheses, as I have already noted. He knew, of course, that these were speculative, that none of them was "metaphysically ultimate fact." But I can find nothing in his writing to suggest that he believed that in principle a decision between these alternatives could not be reached. The task of the natural philosopher (he would have said) was to try to adjudicate between them.

8. As Fine argues in "The Natural Ontological Attitude," this volume.

9. Richard Healey calls it "naive realism"; "naive" not in a deprecatory sense, but as connoting the "natural attitude." See "Quantum Realism: Naiveté Is No Excuse," *Synthese* 42 (1979): 121-144.

10. Especially owing to the developments in recent years of the original quantum formalism, associated not only with physicists (Bell, Kochen, Specker, Wigner) but also with philosophers of science (Cartwright, Fine, Gibbins, Glymour, Putnam, Redhead, Shimony, van Fraassen, and others).

11. This argument may be found, for example, in Fine, "Natural Ontological Attitude," sec. II.

12. G. Chew, "Impasse for the Elementary-Particle Concept," *Great Ideas Today* (Chicago: Encyclopedia Britannica, 1973), 367-389; see 387-389. In his more recent, and very speculative combinatorial topology, Chew has managed to construct a formalism in which the various elementary "particles" are replaced

by combinations of triangles (shades of the *Timaeus*!). Though quarks do not appear in his formalism, Chew has hopes of obtaining all the results that quantum field theory does and perhaps even more.

13. See, for example, W. Heisenberg, "Tradition in Science," in *The Nature of Scientific Discovery*, ed. O. Gingerich (Washington: Smithsonian, 1975), 219-236.

14. In the last few years, this claim has come to seem a lot less plausible, in the short run at least, since quantum field theory has been scoring notable successes, while work on the S-matrix formalism has been all but abandoned.

15. T. Kuhn, *The Structure of Scientific Revolutions*, 2d ed. (Chicago: University of Chicago Press, 1970), 206.

16. See, in particular, L. Laudan, "A Confutation of Convergent Realism," this volume. The quotation is from p. 232.

17. E. Kant, *Critique of Pure Reason*, A226/B273.

18. See G. G. Brittan, *Kant's Theory of Science* (Princeton: Princeton University Press, 1978), chap. 5.

19. B. C. van Fraassen, *The Scientific Image* (Oxford: Clarendon Press, 1980), 73.

20. G. Maxwell, "The Ontological Status of Theoretical Entities," *Minnesota Studies in Philosophy of Science* 3 (1962): 3-27.

21. Van Fraassen, *The Scientific Image*, 16.

22. Ibid., 19.

23. Van Fraassen complicates the picture further by also allowing the sense of 'observable' to depend on the theory being tested. "To find the limits of what is observable in the world described by theory T, we must inquire into T itself, and the theories used as auxiliaries in the testing and application of T." Ibid., 57.

24. Ibid., 16.

25. R. Rorty, *Philosophy and the Mirror of Nature* (Princeton: Princeton University Press, 1979), 274-275.

26. H. Putnam, "What is Realism?" this volume p. 145.

27. See I. Hacking, "Experimentation and Scientific Realism," this volume. It is not clear to me whether one comes up with the same list of entities using Hacking's way as one does with the more usual form of argument relying on explanatory efficacy.

28. Putnam, *Reason, Truth and History*, 57.

29. Ibid., 74.

30. R. Rorty, *Consequences of Pragmatism* (Minneapolis: University of Minnesota Press, 1982), xix.

31. Ibid., xvii.

32. Ibid., xliii.

33. I must say that I have difficulties in seeing that Kant "all but says that he is giving up the correspondence theory of truth" (Putnam, *Reason, Truth and History*, 63), and that he "is best read as proposing for the first time what I have called the 'internalist' or 'internal realist' view of truth" (ibid., 60).

34. Ibid., 55. This puts him close to Dummett's camp in a different philosophical battle.

35. These are briefly sketched in "Realism and Reason," final chapter of H.

Putnam's *Meaning and the Moral Sciences* (London: Routledge, 1978). See also Putnam, "Why There Isn't a Ready-Made World." His main argument is that even if the world did have a "built-in structure" (which he denies), this could not single out *one* correspondence between signs and objects.

36. 'Scientific realism' does not occur in the topic index of Putnam's, *Reason, Truth and History*, even though other 'realisms' are discussed extensively.

37. See Putnam, "Why There Isn't a Ready-Made World."

38. Putnam, *Reason, Truth and History*, 143. It is curious that both he and Rorty (*Consequences of Pragmatism*, xxvi) criticize the realistic tendency to suppose that physics can reach the "one true theory." But they both define the offending sort of realism precisely as the view that supposes that even in the ideal limit such a theory may not be reached. In fact, according to Putnam's own definition, the "one true theory" is, by definition, what physics *does* reach!

39. These become less and less sympathetic as times goes on. I do not see, for example, why a metaphysical realist should defend the claim that "the world consists of some fixed totality of mind-independent objects," or that "there is exactly one true and complete description of the way 'the world is'" (Putnam, *Reason, Truth and History*, 49). Paul Horwich, in an attempt to pin down Putnam's notion, makes it follow from "a more general and fundamental aspect of metaphysical realism," namely, "the view according to which truth is so inexorably separated from our practice of confirmation that we can have no reasonable expectation that our methods of justification are even remotely correct." Horwich claims that Putnam's notion is "committed to an uncomfortable extent to the possibility of unverifiable truth: no truths are verifiable or even inconclusively confirmable" (P. Horwich, "Three Forms of Realism," *Synthese* 51 [1982]: 181-201; see 188, 189). Not only does this go a long way, in my opinion, beyond what Putnam believes metaphysical realism amounts to, but it also makes a straw man of the position. In fact, I know of no philosopher who would defend it in the form in which Horwich states it.

40. Since this was the type of argument that Putnam endorsed in his earlier work, citing Boyd, one can see why he might now have backed away not only from the supporting argument but also from the thesis itself.

41. This is what Horwich calls "epistemological realism." P. Horwich, "Three Forms of Realism," 181. I am not as convinced as he is that this position is "opposed only by the rare skeptic."

42. Fine's essay in this volume appears to fall into this category. The first section of it is devoted to a critique of all the arguments normally brought in support of scientific realism; the second section argues that instrumentalism had a much more salutary influence than realism did on the growth of modern science. But the final section proposes, as the consequence of a "natural ontological attitude," that "there really are molecules and atoms" and rejects the instrumentalist assertion that they are just fictions. But some argument is needed for this, beyond calling this attitude "natural." And to say that the realist adds to this acceptable "core position" an unacceptable "foot-stamping shout of 'Really,'" an "emphasis that all this is really so," leaves me puzzled as to what this difference is supposed to amount to.

43. The issues as to whether these entities *ought* to be attributed privileged

status (as materialism and various forms of reductionism maintain) will not be discussed here.

44. I traced the history and main features of this form of explanation in "Structural Explanation," *American Philosophical Quarterly* 15 (1978): 139-147.

45. M. Simon, *The Matter of Life* (New Haven: Yale University Press, 1971).

46. See the discussion of this in E. McMullin, "The Conception of Science in Galileo's Work," *New Perspectives on Galileo*, ed. R. Butts and J. Pitt (Dordrecht: Reidel, 1978), 209-257.

47. T. Kuhn, *The Essential Tension* (Chicago: University of Chicago Press, 1977), 321-322. See also E. McMullin, "Values in Science," PSA Presidential Address 1982, in *PSA 1982*, vol. 2.

48. For a fuller discussion of the criterion of fertility, see E. McMullin, "The Fertility of Theory and the Unit for Appraisal in Science," *Boston Studies in the Philosophy of Science*, ed. R. S. Cohen et al., 39 (1976): 395-432.

49. See, for instance, P. Wheelwright, *Metaphor and Reality* (Bloomington: Indiana University Press, 1962), esp. chap. 4, "Two Ways of Metaphor"; and E. McMullin, "The Motive for Metaphor," *Proceedings American Catholic Philosophical Association* 55 (1982): 27-39.

50. I have elsewhere developed one instance of this in some detail, the Bohr model of the H-atom as it guided research from 1911 to 1926. See E. McMullin, "What Do Physical Models Tell Us?" in *Logic, Methodology and Philosophy of Science*, Proceedings Third International Congress, ed. B. van Rootselaar (Amsterdam, 1968), 3: 389-396.

51. See, for example, R. Laudan, "The Recent Revolution in Geology and Kuhn's Theory of Scientific Change," in *Paradigms and Revolutions*, ed. G. Gutting (Notre Dame: University of Notre Dame Press, 1980), 284-296; R. Laudan, "The Method of Multiple Working Hypotheses and the Development of Plate-Tectonic Theory," in press.

52. H. Frankel, "The Reception and Acceptance of Continental Drift Theory as a Rational Episode in the History of Science," in *The Reception of Unconventional Science*, ed. S. Mauskopf (Boulder: Westview Press, 1978), 51-89; H. Frankel, "The Career of Continental Drift Theory," *Studies in the History and Philosophy of Science* 10 (1979): 21-66.

53. M. Bradie, "Models, Metaphors and Scientific Realism," *Nature and System* 2 (1980): 3-20.

54. Van Fraassen, *The Scientific Image*, 24.

55. Ibid., 8.

56. H. Putnam, "What is Mathematic Truth?", *Mathematics, Matter and Method* (Cambridge: Cambridge University Press), 69-70.

3

The Current Status of Scientific Realism

Richard N. Boyd

INTRODUCTION

The aim of this essay is to assess the strengths and weaknesses of the various traditional arguments for and against scientific realism. I conclude that the typical realist rebuttals to empiricist or constructivist arguments against realism are, in important ways, inadequate. I diagnose the source of the inadequacies in these arguments as a failure to appreciate the extent to which scientific realism requires the abandonment of central tenets of modern epistemology, and I offer an outline of a defense of scientific realism that avoids the inadequacies in question.

SCIENTIFIC REALISM DEFINED

By 'scientific realism' philosophers typically understand a doctrine which we may think of as embodying four central theses:

1. Theoretical terms in scientific theories (i.e., nonobservational terms) should be thought of as putatively referring expressions; that is, scientific theories should be interpreted "realistically."

2. Scientific theories, interpreted realistically, are confirmable and in fact are often confirmed as approximately true by ordinary scientific evidence interpreted in accordance with ordinary methodological standards.

3. The historical progress of mature sciences is largely a matter of suc-

cessively more accurate approximations to the truth about both observable and unobservable phenomena. Later theories typically build upon the (observational and theoretical) knowledge embodied in previous theories.

4. The reality which scientific theories describe is largely independent of our thoughts or theoretical commitments.

Critics of realism in the empiricist tradition typically deny theses 1 and 2, and qualify their acceptance of 3 so as to avoid commitment to the possibility of theoretical knowledge (however, van Fraassen accepts 1).[1] Antirealists in the constructivist tradition, such as Kuhn,[2] deny 4; however, they may well affirm 1 through 3 on the understanding that the "reality" which scientific theories describe is somehow a social and intellectual construct. As Kuhn and Hanson both argue,[3] a constructivist perspective limits, however, the scope of application of 3, since successive theories can be understood as approximating the truth more closely only when they are part of the same general constructive tradition or paradigm. J. C. C. Smart's version of scientific realism departs from the typical conception in that he rejects 2,[4] holding that distinctively philosophical considerations are required, over and above ordinary standards of scientific evidence, in order to justify our acceptance of the theoretical claims of scientific theories. Since Smart appears to hold that these philosophical considerations are nonevidential, it is perhaps appropriate to treat his position as intermediate between realism and constructivism.

In any event, the principal challenges to scientific realism arise from quite deep epistemological criticisms of 1 through 4. The key antirealist arguments, the standard rebuttals to them in the literature, as well as certain weaknesses in these rebuttals are summarized in the chart in the next section.

ANTIREALISM IN THE EMPIRICIST TRADITION

There is a single, simple, and very powerful epistemological argument that represents the basis for the rejection of scientific realism by philosophers in the empiricist tradition. Suppose that T is a proposed theory of unobservable phenomena, which can be subjected to experimental testing. A theory is said to be empirically equivalent to T just in case it makes the same predictions about observable phenomena that T does. Now, it is always possible, given T, to construct arbitrarily many alternative theories that are empirically equivalent to T but which offer contradictory accounts of the nature of unobservable phenomena. Since scientific evidence for or against a theory consists in the confirmation or

THE BASIC ANTIREALIST ARGUMENTS, THE STANDARD REBUTTALS, AND THEIR WEAKNESSES: A Chart

Antirealist Argument	Standard Rebuttal	Weakness
1. The empiricist argument: empirically equivalent theories are evidentially indistinguishable; therefore, knowledge cannot extend to 'unobservables.'	1.a. There is no sharp distinction between "observables" and "unobservables."	1.a. (i) A sharp distinction can be drawn in a well-motivated way. (ii) In any event, distinction need not be sharp.
	1.b. The empiricist argument ignores the role of auxiliary hypotheses in assessing empirical equivalence.	1.b. The empiricist argument can be reformulated to apply to "total sciences."
	1.c. The "no miracles" argument: If scientific theories weren't (approximately) true, it would be miraculous that they yield such accurate observational predictions.	1.c. It does not address the crucial epistemological claim of the empiricist argument: that since factual knowledge is grounded in experience, it can extend only to observable phenomena.
2. Constructivist arguments:		
2.a. Scientific methodology is so theory-dependent that it is, at best, a construction procedure, not a discovery procedure.	2.a. Pair-wise theory-neutrality of method: for any two rival theories, there are experimental tests based on a method legitimized by both theories.	2.a. It does not address the epistemological point that theory-dependent methodology must be a construction procedure.
2.b. Consecutive "paradigms" in the history of science are not logically commensurable in the way they would be if they embodied theories about a paradigm-independent world.[5]	2.b. It is possible to give an account of continuity of reference for theoretical terms that allows for commensurability of paradigms.	2.b. If the antirealist *epistemological* argument (2.a.) is sound, then such continuity of reference is itself a construct, or at best a matter of continuity of reference to constructs, so that the realist's conception of scientific knowledge of theory-independent reality is still not vindicated.

disconfirmation of one of its observational predictions, *T* and each of the theories empirically equivalent to it will be equally well confirmed or disconfirmed by any possible observational evidence. Therefore, no scientific evidence can bear on the question of which of these theories provides the correct account of unobservable phenomena; at best, it might be possible to confirm or disconfirm the claim that each of these theories is a reliable instrument for the prediction of observable phenomena. Since this construction is possible for any theory *T*, it follows that scientific evidence can never decide the question between theories of unobservable phenomena and, therefore, knowledge of unobservable phenomena is impossible.

This is the central argument of the verificationist tradition. If sound, it refutes scientific realism even if it is not associated with a version of the verifiability theory of meaning. Meaningful or not, theoretical claims are incapable of confirmation or disconfirmation. We may choose the "simplest model" for "pragmatic" reasons, but if evidence in science is experimental evidence, then pragmatic standards for theory choice have nothing to do with truth or knowledge. Scientific realism promises theoretical knowledge of the world, where, at best, it can deliver only formal elegance, or computational convenience.

As I have indicated in the chart, the empiricist argument we have been considering depends on the epistemological principle that empirically equivalent theories are evidentially indistinguishable. The evidential indistinguishability thesis (whether explicit or implicit) represents the key epistemological doctrine of contemporary empiricism and may be thought of as a precise formulation of the traditional empiricist doctrine ("knowledge empiricism" in the phrase of Bennett)[6] that factual knowledge must always be grounded in experiences, and that there is no a priori factual knowledge. (As I shall argue later, the evidential indistinguishability thesis is the wrong formulation of the important epistemological truth in that doctrine; still, it represents the way in which empiricist philosophers of science—and most other empiricists, for that matter—have understood the fundamental doctrine of empiricist epistemology.)

Let us now turn to the standard rebuttals to the antirealist application of the indistinguishability thesis. Perhaps the most commonplace rebuttal to verificationist or empiricist arguments against realism is that the distinction between observable and unobservable phenomena is not a sharp one, and that the fundamental empiricist antirealist argument therefore rests upon an arbitrary distinction.[7] In assessing this rebuttal, it is important to distinguish between the question of the truth of the claim that the distinction between observable and theoretical entities is not sharp, and the question of the appropriateness of this claim as a rebuttal to empiricist antirealism. If scientific realism has somehow been established, then it may well be evident that the distinction in question is epistemologically arbitrary: if we are able to confirm theories of, say, electrons, then we may be able to employ such theories to design electron detecting instruments whose "readings" may have an epistemological status essentially like that of ordinary observations. If, however, it is scientific realism that is in dispute, then the considerations just presented would be inappropriately circular, even if their conclusion is ultimately sound. Only a non-question-begging demonstration that the distinction in question is arbitrary would constitute an adequate rebuttal to the empiricist's strong prima facie case that experimental knowledge cannot extend to the unobservable realm.

If we understand the rebuttal in question in this light, then several responses are available to the empiricist which indicate its weakness as a response to the central epistemological principle of empiricism. In the first place, it is by no means clear that the empiricist need hold that there is a *sharp* distinction between observable and unobservable phenomena in order to show that the distinction is epistemologically nonarbitrary. Suppose that there are entities that represent borderline cases of observability and suppose that there are cases in which it is not clear whether something is being observed or not. Then there will be some entities about which our knowledge will be limited by our capacity to observe them, and there will be cases in which the evidence is equivocal about whether there are entities of a certain sort at all. But the empiricist need hardly resist these conclusions: they are independently plausible, and—provided that there are some clear cases of putative unobservable entities (atoms, elementary particles, magnetic fields, etc.)—the antirealist claims of the empiricist are essentially unaffected.

Moreover, there are at least three ways in which the distinction in question can be made sharper in an epistemologically motivated way. In the first place, there is nothing obviously wrong with the traditional empiricist distinction between sense data and putative external objects. It is often claimed that the failure of logical positivists to construct a sense-datum language shows that the observation-theory dichotomy cannot be formulated in such terms, because it would be impossible to say of a theory that evidence for or against it consists in the confirmation or disconfirmation of observational (that is, sense-datum) predictions that are *deduced* from the theory. Quite so, but the fact remains that some experiences are of the sort we expect on the basis of the acceptance of a given theory, and others are of the sort we would not expect. Whatever the relation of expectation is between theories and sensory experiences, we may define empirical equivalence with respect to it, and affirm the empiricist thesis of the evidential indistinguishability of empirically equivalent theories. The result is *the* classical empiricist formulation of "knowledge empiricism." Insofar as it is plausible, this version of knowledge empiricism provides an argument against scientific realism, even though it also poses the philosophical problem of explicating the relevant expectation relation. In any event, that relation might well be taken to be given by empirical facts about human understanding, rather than by philosophical analysis.

It is true, of course, that the sense-datum formulation of the evidential indistinguishability thesis leads to phenomenalism (at best) about physical objects and other persons. As early logical positivists recognized, this consequence makes it difficult to account for the apparent social and intersubjective character of scientific knowledge. To be sure, this difficulty provides a reason to doubt the truth of the evidential indistinguishability thesis in its sense-datum formulation. But it does not constitute a satisfactory rebuttal to that thesis, nor a satisfactory rebuttal to the antirealist argument we are considering. The sense-datum version of the indistinguishability thesis is, after all, the obvious precise formulation of the doctrine that factual knowledge is always grounded in experience. The empiricist argument against realism is a straightforward application of that thesis. The fact that the thesis in question has inconvenient consequences shows neither that factual knowledge is not grounded in experience, nor that the (sense-datum version of) the indistinguishability thesis is not the appropriate explication of the doctrine that factual knowledge is grounded in this way. Considerations about the public character of science may provide us with reason to think that there must be *something* wrong with the phenomenalist's argument against scientific realism, but these considerations do not provide us with any plausible account of *what* is wrong with it. If I am right, the rebuttal to the sense-datum version of the evidential indistinguishability thesis displays a weakness common to all of the rebuttals to anti-empiricist arguments described in the chart. Each of the principal anti-empiricist arguments raises deep questions in epistemology or semantic theory against scientific realism. The standard rebuttals, insofar as they are effective at all, provide some reason to think that the antirealist arguments in question are unsound, or that realism is true, but they do not succeed in diagnosing the error in these arguments, nor do they point the way to alternative and genuinely realist conceptions of the central issues in epistemology or semantic theory.

It remains to examine the other two ways in which the dichotomy between observable and unobservable phenomena can be sharpened. On the one hand, phenomena might be classed as observable if they are quite plainly observable to persons with normal perceptual abilities. On the other hand, there is the proposal, which seems to be implicit in Maxwell, that entities that may not be directly observable to the unaided senses should count as observable for the purposes of the epistemology of science if they can be detected by the senses when the senses have been aided by devices whose reliability has been previously established by procedures that do not beg the question between empiricists and scientific realists. Roughly at least, the latter proposal can be put this way: Let 0_1 be the class of entities that are observable to the typical unaided senses; for any n, let 0_{n+1} be the class of entities that are detectable by procedures whose legitimacy can be established on the basis of theories that can be established (and can be applied to justify those procedures) without presupposing the existence of entities not in 0_n; the union of the sets 0_n is the class of observables in the sense relevant to the epistemology of science.

Neither of these proposals is without difficulties. Either can be challenged from the perspective of traditional empiricism by a simple application of the sense-datum version of the evidential indistinguishability thesis. The proposal that observability should be defined in terms of what is plainly observable to the unaided senses may be challenged for failing to account, for example, for observations made through a simple light microscope or telescope. The more generous conception is open to the challenge that it fails to see the force of the evidential indistinguishability thesis with respect to its own conception of observability—that it fails, for example, to recognize that there are infinitely many different and evidentially indistinguishable hypotheses that could explain the intersubjectively observable images which are the objective data of light microscopy.

In any event, each of these proposals reflects an important aspect of the intuitive conception that experimental knowledge is grounded in observation. What is important for our purposes is that *either* account of unobservability is sufficient to sustain a significant antirealist application of the evidential indistinguishability thesis. That this is true for the less generous conception of observability is obvious. In regard to the more generous conception, it is important to recognize that what is proposed is not that one may treat as observable whatever phenomena can be identified by "inductive inference to the best explanation" as causes of the results of laboratory "measurement" or "detection."[8] A general appeal to a principle of inductive inference to theoretical explanations would beg the question against the empiricist in this context. Instead, the proposed account of observability depends crucially on the conception that theories whose confirmation by observations is unproblematic from an empiricist point of view can be employed to legitimize an additional level of observables and that this process can then be iterated. The example of light microscopy is illustrative here: the idea is that the lens-makers' equations can be confirmed in a fashion entirely acceptable to empiricism, and that these equations can then be used to legitimize interpreting the images observed through a microscope as images of otherwise unobservable entities.

It is not clear that this approach even gets off the ground as a non-question-begging account of observability. Arguably, the empiricist will hold that the lens-makers' equations, for example, are confirmable only insofar as they are understood to apply to unproblematically observable entities. The application of those equations that underlies the broader conception of observability requires that they be confirmed even when they are understood to apply to the very entities whose observability they are supposed to legitimize. It is by no means clear that objections such as this do not yield the conclusion that $0_n = 0_{n+1}$ for all n.

Even if this problem is somehow circumvented, it is still true that the generous definition of observability is unlikely to legitimize knowledge of the standard "unobservables" which worry the philosopher of science. The reason is this: the account of observability we are considering cannot work to legitimize as observable putative entities that are such that the available procedures for (as a realist would say) measuring and detecting them depend upon explicit theories of those entities themselves, or (worse yet) upon theories of other (putative) entities as well, which are equally unobservable in the traditional sense. In such cases only a question-begging inductive inference to a theoretical explanation of the results of the relevant "measurements" or "detections" would suffice to legitimize the entities in question. But it is almost certain that the basic unobservable putative features of matter (atoms, their constituent particles, electric and magnetic fields, etc.) fall into the category of entities for which legitimization would be question-begging. Therefore the central claims of antirealist empiricism in the philosophy of science will be sustained even if the evidential indistinguishability thesis is so understood as not to rule out the use of, for example, light microscopes in scientific observations.

We may apparently conclude the following about the rebuttal to empiricist antirealist arguments which turns on the claim that the distinction between observable entities and unobservable ones is not sharp and that the empiricist argument therefore rests upon an epistemologically arbitrary distinction: The distinction in question need not be sharp in order to be nonarbitrary. Moreover, there are at least three epistemologically motivated ways of making it sharper. An examination of each of these refinements of the distinction indicates features that might make it reasonable to suppose that there is something problematic about the basic empiricist argument against realism, but none of these considerations provides any diagnosis of the error, nor do any of them allow us to foresee any alternative to the doctrine of evidential indistinguishability of empirically equivalent theories upon which the empiricist argument depends. The standard rebuttals are inadequate in the face of the serious epistemological issues raised by the empiricist position.

I said that we may *apparently* reach these conclusions because it may seem that I have overlooked the real force of the rebuttal under consideration. The real force, it might seem, lies in the following consideration: it has often happened that scientists have postulated unobservable entities and have developed and confirmed, to their satisfaction, theories about them, and that they have much later been able, on the basis of those very theories, to measure or detect those very entities whose existence they earlier had postulated. Examples may include germs, viruses, atoms, and neutrinos. Surely, this shows that the sorts of inductive inference to theoretical explanations in which scientists engage are reliable, whatever empiricists may say.

Taken at face value, this argument is question-begging: it assumes at the outset that what scientific realists describe as "measurement" and "detection" of the entities in question are really measurement and detection. But there is an argument for realism lurking here. It does not turn on the claim that the empiricist has drawn the observable-unobservable dichotomy arbitrarily; such a reading makes the argument question-begging. Instead, what we have is an example of the third anti-empiricist rebuttal (indicated on the chart). In general, that rebuttal points to the astonishing predictive reliability of well-confirmed scientific theories as evidence that they must be approximately true as descriptions of unobservable entities. The cases of predictive reliability that makes this argument plausible are typically those in which predictions quite different from the ones that were involved in the initial confirmation of a theory—and especially predictions arrived at by calculations that take the theoretical machinery of the theory quite seriously—turn out to be surprisingly accurate. In such cases, it seems that miracles are the only alternative to a realist explanation of the success of scientific practice.[9] Cases in which what is predicted are results of (what a realist would call) "measurement" or "detection" of the postulated unobservable entities are especially clear examples of the cases to which this argument applies.

This rebuttal to empiricist antirealism has considerable force (indeed, it is probably the argument that reconstructs the reason why most scientific realists are realists). But it suffers from the same defect that we observed earlier in the case of the first rebuttal: while it provides good reason to think that there must be *something* wrong with the empiricists' argument, it affords us no diagnosis of *what* is wrong with it. No rebuttal to the basic epistemological principle of the empiricist argument (the evidential indistinguishability thesis) flows from this rebuttal; nor is there any rebuttal to the application of that basic principle to the issue of scientific realism. We are provided with a reason to suppose that realism is true, but we are not provided with any epistemology to go with that conclusion.

There remains one rebuttal among the standard responses to empiricist antirealism, and it does seem to challenge directly the evidential indistinguishability thesis. The evidential indistinguishability thesis asserts that empirically equivalent theories are evidentially indistinguishable. But it has been widely recognized by philosophers of science that this is wrong. It might be right, they would argue, if the only predictions from a theory that are appropriate to test are those that can be deduced from the theory in isolation. But it is universally acknowledged that in theory testing we are permitted to use various well-confirmed theories as "auxiliary hypotheses" in the derivation of testable predictions. Thus, two different theories might be empirically equivalent—they might have the same consequences about observable phenomena—but it might be easy to design a crucial experiment for deciding between the theories if one could find a suitable set of auxiliary hypotheses such that when they were brought into play as additional premises, the theories (so expanded) were no longer empirically equivalent.

There is almost no doubt that considerations of this sort rebut any verificationist attempt to classify individual statements or theories as literally meaningful or literally meaningless by the criterion of verifiability in principle. But there is no reason to suppose that the rebuttal based on the role of auxiliary hypotheses is fatal to the basic claim of the evidential indistinguishability thesis, or to its antirealistic application. The reason is this: we may reformulate the evidential indistinguishability thesis so that it applies not to individual theories, but to "total sciences." The thesis, so understood, then asserts that empirically equivalent total sciences are evidentially indistinguishable. Since total sciences are self-contained with respect to auxiliary hypotheses, the rebuttal we have been considering does not apply, and the revised version of the evidential indistinguishability thesis entails that at no point in the history of science could we have knowledge that the theoretical claims of the existing total science are true or approximately true.[10]

One objection that has sometimes been offered against the employment of the notion of a total science is the observation that if, by a "total science," one means the set of well-established theories at a particular time in the history of science, then total sciences are almost certainly always logically inconsistent and, therefore, they have all possible observational consequences and cannot be experimentally confirmed. In this case, as in the case of the objection discussed earlier to the sense-datum version of the evidential indistinguishability thesis, there is an obvious reply. Somehow, scientists manage to cope with inconsistent total sciences; they have a good idea of which tentatively accepted or merely approximate (as they might say) theories should not be employed together in making predictions. They have a pretty good idea which predictions not to trust. All we need to do is to define empirical equivalence with respect to the practice of scientists. The evidential indistinguishability thesis formulated with respect to total sciences in this way yields the antirealist conclusion of empiricists, and it certainly seems reasonable to hold that some such version of the evidential indistinguishability thesis represents the obvious interpretation of "knowledge empiricism" once the role of auxiliary hypotheses is acknowledged. Thus, the fact that auxiliary hypotheses play a crucial role in theory confirmation does not constitute a significant rebuttal to a sophisticated version of the standard empiricist argument against scientific realism.

There *is* a point regarding the use of auxiliary hypotheses that can be made the basis for a very strong defense of scientific realism. The use of auxiliary hypotheses, like other applications of what positivists called the "unity of science" principle, depends upon judgments of univocality regarding different occurrences of the same theoretical terms. It is possible to argue that only a realist conception of the semantics and epistemology of science can account for the role of such univocality judgments in contributing to the reliability of scientific methodology.[11] But this argument is not anticipated in the standard rebuttals to empiricist antirealism.

One must conclude that the standard rebuttals to the central empiricist argument against scientific realism are significantly flawed. Where they do provide reason to suspect that the empiricist argument is unsound (or, more directly, that realism is true), they do not provide any effective rebuttal to the main epistemological principle (the evidential indistinguishability thesis) upon which the empiricist argument depends, nor do they indicate respects in which the application of that principle to the question of realism is unwarranted.

CONSTRUCTIVIST ANTIREALISM

There is a single basic empiricist argument against realism, and it is an argument of striking simplicity and power. In the case of constructivist antirealism, the situation is much more complex. In part, at least, this is

so because constructivist philosophers of science have typically been led
to antirealist conclusions by reflections upon the results of *detailed* exam-
inations of the history and actual methodological practices of science as
well as by reflections on the psychology of scientific understanding. Different philosophers have focused on different aspects of the complex procedures of actual science as a basis for antirealist conclusions. Nevertheless, it is possible, I believe, to identify the common thread in all of these
diverse arguments. Roughly, the constructivist antirealist reasons as
follows: The actual methodology of science is profoundly theory-dependent. What scientists count as an acceptable theory, what they count as
an observation, which experiments they take to be well designed, which
measurement procedures they consider legitimate, what problems they
seek to solve, and what sorts of evidence they require before accepting a
theory—which are all features of scientific methodology—are in practice
determined by the theroetical tradition within which scientists work.
What sort of world must there be, the constructivist asks, for this sort of
theory-dependent methodology to constitute a vehicle for gaining knowledge? The answer, according to the constructivist, is that the world that
scientists study, in some robust sense must be defined or constituted by
or "constructed" from the theoretical tradition in which the scientific
community in question works. If the world that scientists study were not
partly constituted by their theoretical tradition, then, so the argument
goes, there would be no way of explaining why the theory-dependent
methods that scientists use are a way of finding out what is true.

To this argument there is typically added another which addresses an
apparent problem with constructivism. The problem is that scientists
seem sometimes to be forced by new data to abandon important features
of their current theories and to adopt radically new theories in their
place. This phenomenon, it would seem, must be an example of scientific
theories being brought into conformity with a theory-independent world,
rather than an example of the construction of reality within a theoretical
tradition. In response to this problem, constructivism often asserts that
successive theories in science that represent the sort of radical "breaks" in
tradition at issue are "incommensurable."[12] The idea here is that the standards of evidence, interpretation, and understanding dictated by the old
theory, on the one hand, and by the new theory, on the other hand, are
so different that the transition between them cannot be interpreted as
having been dictated by any common standards of rationality. Since
there are no significant theory-independent standards of rationality, it
follows that the transition in question is not a matter of rationally adopting a new conception of (theory-independent) reality in the light of new
evidence; instead, what is involved is the adoption of a wholly new conception of the world, complete with its own distinctive standards of

rationality. In its most influential version, this argument incorporates the claim that the semantics of the two consecutive theories change to such an extent that those terms that they have in common should not be thought of as having the same referents in the two theories.[13] Thus, transitions of the sort we are discussing ("scientific revolutions" in Kuhn's terminology) involve a total change of theoretical subject matter.

There are two closely related standard rebuttals to these antirealist arguments. In the first place, against the claim that realism must be abandoned because scientific methodology is too theory-dependent to constitute a discovery (as opposed to a construction) procedure, it is often replied that for any two rival scientific theories it is always possible to find a methodology for testing them that is neutral with respect to the theories in question. Thus the choice between rival scientific theories on the basis of experimental evidence can be rational even though experimental methodology is theory-dependent. The outcome of a "crucial experiment" that pits one rival theory against another need not be biased, since such an experiment can be conducted on the basis of a methodology that—however theory-dependent—is not committed to either of the two contesting theories.

Against the incommensurability claim, it is often argued that an account of reference for theoretical expressions can be provided that makes it possible to describe scientific revolutions as involving continuity in reference for the theoretical terms common to the laws of the earlier and later theoretical traditions or "paradigms." With such referential continuity comes a kind of continuity of methodology as well, because (assuming continuity of reference) the actual cases of scientific revolutions typically result in the preservation of some of the theoretical machinery of the earlier paradigm in the structure of the new one, and this, in turn, guarantees a continuity of methodology.

Neither of these rebuttals is fully adequate as a response to constructivist antirealism. Consider first the claim that for any two rival theories there is a methodology for testing them that is neutral with respect to the issues on which they differ ("pair-wise theory-neutrality of method" in the chart). It is generally true that for theoretical rivalries that arise in actual science, a relevantly neutral testing methodology will exist. Indeed, the use of such "neutral" testing methodologies is a routine part of what Kuhn calls "normal science." And indeed, the existence of such methodologies helps to explain how scientists can appeal to common standards of rationality even when they have theoretical differences of the sort that influence methodological judgments. Nevertheless, pair-wise theory-neutrality of method does not provide a reason to reject the antirealist conclusions of the constructivist.

Remember that what the constructivist argues is that a general methodology that is predicated upon a particular theoretical tradition and is theory-determined to its core cannot be understood as a methodology for discovering features of a world that is not in some significant way defined by that tradition. All that the doctrine of pair-wise theory-neutral methods asserts is that within the theoretical and methodological tradition in question, there are available experimental procedures that are neutral with respect to quite particular disputes between alternative ways of modifying or extending that very tradition. There is no suggestion of a procedure by which scientific methodology can escape from the presuppositions of the tradition and examine objectively the structure of a theory-independent world. Insofar as the profound theory-dependence of method raises an epistemological problem for realism, the pair-wise theory-neutrality of methods does not provide an answer to it.

Perhaps surprisingly, it does not help either to demonstrate that successive paradigms are commensurable. Suppose that a satisfactory account of referential continuity for theoretical terms during scientific revolutions is available.[14] Suppose further (what is not implied by the former claim) that the theoretical continuity thus established during revolutionary periods is such that the transition between the prerevolutionary theory and the postrevolutionary one is governed by a continuously evolving standard of scientific rationality. If these suppositions are true, then much of what Kuhn, for example, has claimed about the history of science will be mistaken: postrevolutionary scientists will (contrary to Kuhn) be building on the theoretical achievements of their prerevolutionary predecessors; the adoption of new "paradigms" will be scientifically rational; and it will not involve a "Gestalt shift" in the scientific community's understanding of the world, whatever may be the case for some individual scientists. *But*, the basic constructivist epistemological objection to scientific realism will still be unrebutted. If the theory-dependence of methodology provides reason to doubt that scientific inquiry possesses the right sort of objectivity for the study of a theory-independent world, then the sort of historical continuity through scientific revolutions we are considering will not address that doubt. Only if the transitional methodology during revolutions were largely theory-neutral would the fact of methodological and semantic continuity between revolutions provide, by itself, a rebuttal to the constructivist antirealist; but there is no chance that such theory-independence could be demonstrated by the sort of rebuttal to incommensurability we are considering. Indeed, there is no reason of any sort to suppose that such a theory-neutral method ever prevails.

In the present case, as in the case of the standard rebuttals to empiricist antirealism, it is by no means true that the standard rebuttals to the constructivist arguments are irrelevant to the issue of scientific realism. If there were no such phenomenon as pair-wise theory-neutrality of method, then it would be hard to see how there could be any sort of scientific objectivity, realist or constructivist. If there is no way of defending the continuity of subject matter and methodology during most of the episodes which Kuhn calls scientific revolutions, then the realist conception of science is rendered most implausible. The point is that, even though these prorealist rebuttals to constructivist antirealism do provide some support for aspects of the realist position, they fail to offer any reason to reject the basic epistemological argument against realism which the constructivist offers.

EMPIRICISM AND CONSTRUCTIVISM

Kuhn presents his constructivist account of science as an alternative to the tradition of logical empiricism and, indeed, there is much he says with which traditional positivists would disagree. There are, nevertheless, important similarities between the constructivist and the empiricist approach to the philosophy of science. Kuhn, for example, relies on the late positivist 'law-cluster' account of the meaning of theoretical terms in his famous argument against the semantic commensurability of successive paradigms.[15] Similarly, R. Carnap's mature positivism of the early 1950s has much in common with Kuhn's views. In particular, "Empiricism, Semantics, and Ontology"[16] offers an account of the criteria for the rational acceptance of a linguistic framework which is surprisingly like a formalized version of Kuhn's view.[17] We may say with some precision what the points of similarity between Kuhn and Carnap are. In the first place, they are agreed that the day-to-day business of the development and testing of scientific theories is governed by broader and more basic theoretical principles, including the most basic laws and definitions of the relevant sciences.

There is a far deeper point of agreement. Kuhn, and constructivists generally, cannot consistently accept the principle of the evidential indistinguishability of empirically equivalent total sciences; they hold, after all, that 'facts'—insofar as they are the subject matter of the sciences—are partly constituted or defined by the adoption of "paradigms" or theoretical traditions, so that there is a sort of a priori character to the scientist's knowledge of the fundamental laws in the relevant tradition or paradigm. But they agree with logical empiricists in holding that any rational constraint on theory acceptance that is not purely pragmatic and that does not accord with the evidential indistinguishability thesis must be essen-

tially conventional. For Carnap and other positivists, the conventions
are essentially linguistic: they amount to the conventional adoption of
one set of "L-truths" rather than another. For Kuhn and other constructivists, the conventions go far deeper: they amount to the social construction of reality and of experimental "facts." What neither empiricists nor
constructivists accept is the idea that the regulation of theory acceptance
by features (linguistic or otherwise) of the existing theoretical tradition
can be a reliable guide to the discovery of theory-independent matters of
fact.

Of course, empiricists and constructivists differ, especially regarding
the extent to which experimental observations can be divorced from
theoretical considerations, and (if constructivists are "relativists" in the
Kuhnian tradition) about the methodological commensurability of successive theoretical traditions or paradigms. It is interesting to note that
Kuhn and the Carnap of the early 1950s do not disagree about the
semantic incommensurability of the theoretical portions of alternative
linguistic frameworks for science—neither accepts any doctrine of continuity of *reference* for theoretical terms in the transition to alternative
frameworks. Indeed, for Carnap, questions of reference and ontology are
meaningless when raised outside the scope of some particular linguistic
framework. That Kuhn and Carnap should agree to this extent about the
semantics of theoretical terms is less surprising when one realizes that
Kuhn's account of the meaning of such terms is simply a more subtle and
historically more accurate version of Carnap's.[18]

One further point of agreement between empiricists and constructivists
is significant for our purposes. Empiricist philosophers of science deny
that knowledge of theoretical entities is possible. But it is no part of contemporary empiricism to deny that the scientific method yields objective
instrumental knowledge: knowledge of regularities in the behavior of
observable phenomena. It is important to see that this point is not seriously contested by constructivist philosophers of science. It is true that
constructivists insist that observation in science is significantly theory-
determined, and that Kuhn, for example, emphasizes that experimental
results that are anomalous in the light of the prevailing theoretical conceptions are typically ignored if they cannot readily be assimilated into
the received theoretical framework. But no serious constructivist maintains that the predictive reliability of theories in mature science or the
reliability of scientific methodology in identifying predictively reliable
theories is largely an artifact of the tendency to ignore anomalous results.
Such a view would be nonsensical in the light of the contributions of pure
science to technological advance.

There is one point that, whether ultimately compatible with empiricism or not, is certainly emphasized by constructivists much more than by empiricists, and is especially relevant when one considers the role of scientific methodology in producing instrumental knowledge. It was recognized early on by logical empiricists that any account of the methodology of science requires some account of the way in which the "degree of confirmation" of a theory, given a body of observational evidence, is to be determined. More recently, N. Goodman has, following Locke, raised a question that is really a special case of the problem of determining degree of confirmation.[19] Any account of the methodology of science must account for judgments of 'projectability' of predicates or, to put the issue more broadly, it must provide an account of the standards by which scientists determine which general conclusions are even real candidates for acceptance given an (always finite) body of available data.[20] This question is interesting precisely because, given any finite body of data, there are infinitely many different general theories that are logically consistent with those data (indeed, there will be infinitely many such theories that are pair-wise empirically inequivalent, given the existing total science as a source of auxiliary hypotheses).

What Kuhn and other constructivists insist (correctly, I believe) is that judgments of projectability and of degrees of confirmation are quite profoundly dependent upon the theories that make up the existing theoretical tradition or paradigm. The theoretical tradition dictates the terms in which questions are posed and the terms in which possible answers are articulated. In a similar way, theoretical considerations dictate the standards for experimental design and for the assessment of the experimental evidence. Assuming this to be true, and assuming, as reasonable constructivists must, that the reliability of scientific methodology in producing instrumental knowledge is not to be explained largely by the tendency to ignore anomalous data, we can see that an important epistemological issue emerges regarding judgments of projectability and of degree of confirmation: why should so theory-dependent a methodology be reliable at producing knowledge about (largely theory-independent) observable phenomena?

A related question about what we might call the "instrumental reliability" of scientific method should prove challenging both to Kuhn, and to empiricists who share with Kuhn the law-cluster theory of the meaning of theoretical terms. Judgment of univocality for particular occurrences of the (lexicographically) same theoretical term play an important epistemological role in scientific methodology. This is evident since commonplaces such as the use of auxiliary hypotheses in theory testing, or applications of the principle of "unity of science" in the derivation of observational predictions from theories that have already been accepted, depend

upon prior assessments of univocality. This means that scientific standards for the assessment of univocality for token occurrences of theoretical terms must play a crucial epistemological role, and it must be the business of an adequate account of the language of science to say what those
standards are *and* why they are such as to render instrumentally reliable
the methodological principles in actual science which depend upon univocality judgments.[21]

Unlike earlier positivist theories of meaning for theoretical terms (like
operationalism, for example) the law-cluster theory does not say what it
is for two tokens of orthographically the same theoretical term to occur
with the same meaning or reference. The meaning of a theoretical term is
given by the most basic laws in which it occurs; this may possibly tell us
something about diachronic questions about univocality of theoretical
terms. But suppose that t and t' are two tokens of orthographically the
same theoretical term, used at the same time, and that neither t nor t'
occurs in a law that is fundamental in the sense relevant to the law-cluster
theory. This latter condition describes the circumstances of almost all
tokens of theoretical terms in actual scientific usage. Under the circumstances in question, the law-cluster theory says nothing about the question of whether t and t' have the same meaning or reference. Only when
the synchronic problem of univocality in such cases is presumed to have
already been solved does the law-cluster theory have anything to say
about univocality for theoretical terms. The law-cluster theory is thus
entirely without the resources to address the important question of the
contribution that judgments of univocality for theoretical terms make to
the instrumental reliability of scientific methodology.

Thus we have identified two questions that pose especially sharp challenges to both empiricist and constructivist conceptions of science: why
are theory-dependent standards for assessing projectability and degrees
of confirmation instrumentally reliable? and, how do judgments of univocality for theoretical terms contribute to the instrumental reliability of
scientific methodology? I shall argue in the next section that answers to
these challenges provide the basis for a new and more effective defense of
scientific realism.

DEFENDING SCIENTIFIC REALISM

Elsewhere, I have offered a defense of scientific realism against empiricist
antirealism which proceeds by proposing that a realistic account of scientific theories is a component in the only scientifically plausible explanation for the instrumental reliability of scientific methodology.[22] What I

propose to do here is to summarize this defense very briefly and to indicate how it also constitutes a defense of scientific realism against constructivist criticisms, and how it avoids the weaknesses in the traditional rebuttals to antirealist arguments.

The proposal that scientific realism might be required in order to explain adequately the instrumental reliability of scientific methodology can be motivated by reexamining the principal constructivist argument against scientific realism (see 2.*a* in the chart). The constructivist asks, What must the world be like in order that a methodology so theory-dependent as ours could constitute a way of finding out what is true? She answers: The world would have to be largely defined or constituted by the theoretical tradition that defines that methodology. It is clear that another answer is at least possible: the world might be one in which the laws and theories embodied in our actual theoretical tradition are approximately true. In that case, the methodology of science might progress dialectically. Our methodology, based on approximately true theories, would be a reliable guide to the discovery of new results and the improvement of older theories. The resulting improvement in our knowledge of the world would result in a still more reliable methodology leading to still more accurate theories, and so on.[23]

What I have argued in the works cited above is that this conception of the enterprise of science provides the only scientifically plausible explanation for the instrumental reliability of the scientific method. In particular, I argue that the reliability of theory-dependent judgments of projectability and degrees of confirmation can only be satisfactorily explained on the assumption that the theoretical claims embodied in the background theories which determine those judgments are relevantly approximately true, and that scientific methodology acts dialectically so as to produce in the long run an increasingly accurate theoretical picture of the world. Since logical empiricists accept the instrumental reliability of actual scientific methodology, this defense of realism represents a cogent challenge to logical empiricist antirealism. It remains to be seen whether it has the weaknesses of more traditional responses to empiricist antirealism, but, first, let us examine its relevance to constructivism.

First, it should be observed that the argument for realism that I have indicated is a direct response to the central constructivist argument against realism. If the argument for realism is correct, then we can see *what* is wrong with the central constructivist argument: the constructivist's epistemological challenge to scientific realism rests upon the wrong explanation for the reliability of the scientific method as a guide to truth.

It is equally important to see that there is no answer within a purely constructivist framework to the question of why the methods of science are *instrumentally* reliable. The instrumental reliability of particular scientific theories cannot be an artifact of the social construction of reality. Even within "pure" science this is acknowledged, for example, by Kuhn. The anomalous observations that (sometimes) give rise to "scientific revolutions" cannot be reflections of a fully paradigm-dependent world: anomalies are defined as observations that are inexplicable within the relevant paradigm. It is even more evident that theory-dependent technological progress (the most striking example of the instrumental reliability of scientific *methods* as well as theories) cannot be explained by an appeal to social construction of reality. It cannot be that the explanation for the fact that airplanes, whose design rests upon enormously sophisticated theory, do not often crash is that the paradigm *defines* the concept of an airplane in terms of crash resistance. If the empiricist cannot offer a satisfactory account of the instrumental reliability of scientific method (as I have argued in the works cited), then the constructivist—who even more than the empiricist emphasizes the theory dependence of that method—cannot do so either. Thus, the epistemological thrust of constructivism is directly challenged by the argument for scientific realism under consideration.

It is clear, moreover, that if scientific realism is defended in this way, then the more traditional rebuttals to constructivist antirealism are rendered fully effective. If the fundamental epistemological thrust of constructivism is mistaken, then (as I indicated earlier) the pair-wise theory-neutrality of scientific methodology, and the continuity of reference of theoretical terms and methods across "revolutions" are crucial components in the defense of scientific realism.

Let us turn now to the question of whether the defense of realism we are considering has the weakness of the more traditional rebuttals to empiricist antirealism. Those rebuttals had the defect that, while they provided some reason to believe that scientific realism is true, they offered no insight into the question of what is wrong with the crucial empiricist argument against realism. Here, the argument under consideration succeeds where the more traditional arguments fail. What is wrong with the fundamental empiricist argument is that the principle that empirically equivalent total sciences are evidentially indistinguishable is false, and it represents the wrong reconstruction of the perfectly true doctrine that factual knowledge is grounded in observation.

The point here is that if the realist and dialectical conception of scientific methodology is right, then considerations of the theoretical plausibility of a proposed theory in the light of the *actual* (and approximately true) theoretical tradition are *evidential* considerations: results of such assessments of plausibility constitute evidence for or against proposed theories. Indeed, such considerations are a matter of theory-mediated empirical evidence, since the background theories, with respect to which assessments of plausibility are made, are themselves empirically tested (again, in a theory-mediated way). Theory-mediated evidence of this sort is no less empirical than more direct experimental evidence—largely because the evidential standards that apply to so-called direct experimental tests of theories are theory-determined in just the same way that judgments of plausibility are. In consequence, the *actual* theoretical tradition has an epistemically privileged position in the assessment of *empirical* evidence. Thus, a total science whose theoretical conception is significantly in conflict with the received theoretical tradition is, for that reason, subject to "indirect" but perfectly real prima facie disconfirmation relative to an empirically equivalent total science that reflects the existing tradition. The evidential indistinguishability thesis is therefore false, and the basic empiricist antirealist argument is fully rebutted.[24]

It might seem that the realist conception that theoretical considerations in science are evidential would reflect a weakening of ordinary standards of evidential rigor in science. After all, on the realist conception, a theory can get evidential support both from (direct) experimental evidence and from (indirect) theoretical considerations. Moreover, the realist proposal might seem to make it impossible to disconfirm traditional theories, treating them as a priori truths in much the same way that the constructivist conception does. Neither of these claims proves to be sound. In the first place, rigorous assessment of experimental evidence in science depends fundamentally upon just the principle that theoretical considerations are evidential: that is why a realist conception of theories is necessary to account for the instrumental reliability of our standards for assessing experimental evidence. Second, the realist conception of theory-mediated experimental evidence does not have the consequence that any traditional laws are immune from refutation. Instead, it provides the explanation of how rigorous testing of these and other laws is possible. The dialectical process of improvement in the theoretical tradition does not preclude, but instead requires, that particular laws or principles in the tradition may have to be abandoned in the light of new evidence.

Let us turn now to the second puzzle about the instrumental reliability of scientific method which was raised at the end of the preceding section: how to account for the epistemic reliability of judgments of univocality for theoretical terms. The realistic account of the instrumental reliability of judgments of projectability requires that the kinds or categories into which features of the world are sorted for the purpose of inductive inference be determined by theoretical considerations rather than being fixed by conventional definitions, however abstract.[25] In particular, the law-cluster theory of meaning, understood conventionally, is inadequate as an account of the "definitions" of theoretical terms in science. It has been widely recognized that if theoretical terms in science are to refer to entities or kinds whose "essences" are determined by empirical investigation rather than by stipulation, then the traditional conception of reference fixing by stipulatory conventions must be abandoned for such terms in favor of some "causal" or "naturalistic" theory of reference.[26]

Given the distinctly realistic conception of scientific knowledge described previously, it is possible to offer a naturalistic theory of reference which is especially appropriate to an understanding of the role of theoretical considerations in scientific reasoning. Such a theory defines reference in terms of relations of "epistemic access."[27] Roughly, a (type) term t refers to some entity e just in case complex causal interactions between features of the world and human social practices bring it about that what is said of t is, generally speaking and over time, reliably regulated by the real properties of e. Because such regulation of what we say by the real features of the world depends upon the approximate truth of background theories, the approximate reliability of measurement and detection procedures, and the like, the epistemic access account of reference can explain the grains of truth in such previous accounts of reference as the law-cluster theory, or operationalism.[28]

Consider now the question of univocality for two token occurrences of orthographically the same theoretical term. Such a pair of terms will be coreferential just in case the social history of each of their occurrences links them, by the relevant sort of causal relations, to a situation of reliable belief regulation by the actual properties of the same feature of the world. The relevant sorts of causal relations are to be determined by epistemology, construed as an empirical investigation into the mechanisms of reliable belief regulation. Thus it is an empirical question, not a conceptual one, whether two such tokens are univocal.

Because the epistemic access account of reference can explain the grains of truth in the other theories of reference for theoretical terms which have been advanced to explain the actual judgments of scientists and historians about issues of univocality, there is every reason to believe that the epistemic access account can explain why the ordinary standards for judging univocality that prevail in science are reliable indicators of actual coreferentiality. Together with the realist's conception that scientific methodology produces (typically and over time) approximately true beliefs about theoretical entities, the epistemic access account of reference provides an explanation of how univocality judgments contribute to the reliability of scientific methodology, an explanation that is fully in accord with the general realist conception of scientific methodology described here.

Finally, the epistemic access account provides a precise formulation of the crucial realist claim that typically (perhaps despite changes in law-clusters) there is continuity of reference across "scientific revolutions."[29] Indeed, it permits us to integrate cases of what H. Field calls "partial denotation"[30] into a general theory of reference and thus to treat cases of "denotational refinement" as establishing referential continuity in the relevant sense.

If the dialectical and realistic conception of scientific methodology described here and the related epistemic access conception of reference are approximately correct, then together they constitute a rebuttal to both empiricist and constructivist antirealism which suffers none of the shortcomings of the more traditional rebuttals, while at the same time accommodating the insights that the more traditional rebuttals provide.

SCIENTIFIC REALISM AND METAPHILOSOPHY

I have examined traditional rebuttals to antirealist arguments in the empiricist and constructivist traditions and have suggested that these rebuttals have the weakness that they do not provide a diagnosis of the epistemological errors that must—if realism is true—lie behind the standard argument against realism. I indicated how a distinctly realistic and dialectical conception of scientific methodology, together with a closely related naturalistic conception of reference, could provide the basis for a defense of realism that does diagnose the epistemological errors in antirealist arguments. If the conception of scientific knowledge and language described here is correct, then it has implications for philosophical methodology which are sufficiently startling that they may help to explain why the dialectical and realist account of the reliability of scientific methodology was not put forward earlier as the epistemological foundation for scientific realism.

I believe it is fair to say that scientific realists have had a conception of their dispute with empiricist and (more recently) with constructivist antirealists according to which they shared with their opponents a general conception of the logic and methods of science, and according to which the dispute between realists and antirealists was over whether that logic and those methods were adequate to secure theoretical knowledge of a theory-independent reality. It was not anticipated that a new and distinctly realist general account of the methods of science would be necessary in order to defend scientific realism. This conception of a shared account of the logic and methods of science was advanced explicitly by E. Nagel, in discussing the realism-empiricist dispute:

It is difficult to escape the conclusion that when the two opposing views on the cognitive status of theories are stated with some circumspection, each can assimilate into its formulation not only the facts concerning the primary subject matter explored by experimental inquiry but also the relevant facts concerning the logic and procedures of science. In brief, the opposition between these views is a conflict over preferred mode of speech.[31]

It is evident that the argument for scientific realism described in the preceding section departs from this understanding. According to that argument, no empiricist or constructivist account of the methods of science can explain the phenomenon of instrumental knowledge in science, the very kind of scientific knowledge about which realists, empiricists, and constructivists largely agree. Only on a distinctly realist conception of the logic and methods of science—a conception that empiricists and constructivists cannot share—can instrumental knowledge be explained.

The distinctly realist conception of the methodology of science departs even farther from the normal conception of the epistemology of science. At least since Descartes, the characteristic conception of epistemology in general has been that the most basic epistemological principles—the basic canons of reasoning or justification—should be defensible a priori. Thus, for example, almost all empiricists have thought that "knowledge empiricism" represented an a priori truth about knowledge, and that the most basic principles of inductive reasoning, whatever they are, can be defended a priori. Similar conceptions are even more clearly seen in the rationalist and Kantian traditions. What is striking is that, if the distinctly realist account of scientific knowledge is sound, then the most basic principles of inductive inference lack any a priori justification. That this is so can be seen by reflecting on what the scientific realist must say about the history of the scientific method.

According to the distinctly realist account of scientific knowledge, the reliability of the scientific method as a guide to (approximate) truth is to be explained only on the assumption that the theoretical tradition that defines our actual methodological principles reflects an approximately true account of the natural world. On that assumption, scientific methods will lead to successively more accurate theories and to successively more reliable methodological practices.[32] If we now inquire how the theoretical tradition came to embody sufficiently accurate theories in the first place, the scientific realist cannot appeal to the scientific method as an explanation, because that method is epistemically reliable only on the assumption that the relevant theoretical tradition already embodies a sufficiently good approximation to the truth. The realist, as I have portrayed here, must hold that the reliability of the scientific method rests upon the logically, epistemically, and historically contingent emergence of suitably approximately true theories. Like the causal theorist of perception or other "naturalistic" epistemologists, the scientific realist must deny that the most basic principles of inductive inference or justification are defensible a priori. In a word, the scientific realist must see epistemology as an *empirical* science.[33]

Closely analogous consequences follow from the epistemic access account of reference when it is applied in the light of scientific realism. The question of whether two tokens of a theoretical term are coreferential is, for example, a purely empirical question that cannot be resolved by conceptual analysis. If we think of the "meaning" of a theoretical term as comprising those features of its use in virtue of which it has whatever referent it in fact has, then meanings of theoretical terms are not given by a priori stipulations or social conventions. It is logically, historically, and epistemically contingent matter which features of the use of a given term constitute its meaning, in the sense of meaning relevant to referential semantics. There simply are not going to be any important analytic or conceptual truths about any scientifically interesting subject matter.

If these controversial consequences of a thoroughgoing realist conception of scientific knowledge are sound, then it would be hard to escape a still more controversial conclusion: philosophy is itself a sort of empirical science. It may well be a normative science—epistemology, for example, may aim at understanding which belief-regulating mechanisms are reliable guides to the truth—but it will be no less an empirical science for being normative in this way.

ISSUES OF PHILOSOPHICAL METHOD

In this section, I shall discuss two issues of philosophical methodology raised by the arguments for scientific realism described in the section "Defending Scientific Realism." First, I shall discuss, at some length, an important challenge raised by Arthur Fine against the basic strategy of those arguments. Then, I shall discuss, somewhat more briefly, certain questions about the ways in which evidence from the history of science bears upon the arguments in question.

THE CHALLENGE TO ABDUCTION

Fine raises a number of interesting objections to the arguments for scientific realism outlined earlier.[34] Of these objections, one is particularly striking because it challenges not the details of the argument for realism, but its basic philosophical strategy.

Fine's objection is extremely simple and elegant. The proposed defense of realism proceeds by an abductive argument: we are encouraged to accept realism because, realists maintain, realism provides the best explanation of the instrumental reliability of scientific methodology. Suppose for the sake of argument that this is true. We are still not justified in believing that realism is true. This is so because the issue between realists and empiricists is precisely over the question of whether or not abduction is an epistemologically justifiable inferential principle, especially when, as in the present case, the explanation postulated involves the operation of unobservable mechanisms. After all, if abductive inference is justifiable, then there is no epistemological problem about the theoretical postulation of unobservables in the first place. It is precisely abductive inference to unobservables that the standard empiricist arguments call into question. Thus, the abductive defense of realism we are considering is viciously circular.

It is reasonable to think of Fine's objection in the light of the previous discussion of the "no miracles" argument for realism (discussed in the section "Antirealism in the Empiricist Tradition"). Against the "no miracles" argument I argued that, even if realism provides the best explanation for the predictive reliability of scientific theories, there remains for the realist the problem that this fact does not constitute a rebuttal to the very powerful epistemological considerations that form the basis for empiricist antirealism. Fine, in effect, presents a generalized version of this response to the "no miracles" argument. In the first place, Fine's version of the response in question applies not only to the "no miracles" argument but to any argument for realism that adduces realism as (a component of) the best explanation for some natural phenomenon. In particular, Fine's objection applies to the argument for realism offered in the section on "Defending Scientific Realism." Suppose now that scientific realism provides the best explanation for the reliability (not just of individual theories but) of the methodology of science as a whole. This fact *by itself* does not constitute a rebuttal to the epistemological principles upon which the empiricist criticism of realism rests.

Moreover, Fine's objection diagnoses not only a weakness in such arguments for realism but a circularity as well. The issue of scientific realism is—at least insofar as the dispute between realists and empiricists is concerned—a debate over the legitimacy of inductive inference to the best explanation, at least in those cases in which the explanation in question postulates unobservable entities. Arguments for realism of the sort which Fine criticizes employ just this sort of inference, and, thus, simply beg the question between realists and empiricist antirealists.

Several things must be said in reply to Fine's subtle and elegant objection. In the first place, Fine's entirely correct insistence that the issue between empiricists and realists is over the legitimacy of abductive inference is a double-edged sword. While it facilitates the identification of a sort of circularity in arguments for realism, it also highlights the epistemological oddity of consistent empiricism. The rejection of abduction or inference to the best explanation would place quite remarkable strictures on intellectual inquiry. In particular, it is by no means clear that students of the sciences, whether philosophers or historians, would have any methodology left if abduction were abandoned. If the fact that a theory provides the best available explanation for some important phenomenon is not a justification for believing that the theory is at least approximately true, then it is hard to see how intellectual inquiry could proceed. Of course, the antirealist might accept abductive inferences whenever their conclusions do not postulate unobservables, while rejecting such inferences to "theoretical" conclusions. In this case, however, the burden of proof would no longer lie exclusively on the realist's side: the antirealist must justify the proposed limitation on an otherwise legitimate principle of inductive inference.

This difficulty for the antirealist is exacerbated when one considers the issue of inductive inference in science itself. It must be remembered that empiricist philosophers of science do not intend to be fully skeptical: it is no part of standard empiricist philosophy of science to reject all nondeductive inferences. Instead, a selective skepticism is intended: (some) inductive generalizations about observables are to be epistemologically legitimate, while inferences to conclusions about unobservables are to be rejected. As Hanson, Kuhn, and others have shown, the actual methods of science are profoundly theory-dependent. I have emphasized in previously cited publications that this theory-dependence extends to the methods scientists employ in making inductive generalizations about observable phenomena. Both the choice of generalizations that are seriously advanced and the assessment of the evidence for or against them rest upon theoretical inferences that manifest, or depend upon, the sort of abductive inferences to which the empiricist objects. In the terminology of recent empiricism, both the assessment of "projectability" of predicates, and the assessment of "degree of confirmation" of generalizations about observables depend in practice, upon inferences about "theoretical entities." Of course, acknowledging these facts about scientific practice would not commit the empiricist to agreeing that realism provides the best explanation for the instrumental reliability of scientific methodology nor, as Fine insists, would agreeing to that proposition commit the empiricist to holding that there is any reason to believe that realism is true. Nevertheless it certainly seems that, unless—as is very unlikely—the apparent theory-dependence of inductive inference about observables is really only apparent, the empiricist who rejects abductive inferences regarding unobservables must hold that even the inductive inferences scientists make about observables are unjustified.

It might seem that there is an easy way out of this last difficulty for the empiricist. Suppose that inductive inferences about observables in science are genuinely theory-dependent and that, therefore, the (necessarily theoretical) justifications, which scientists would ordinarily offer in defense of their inductive inferences about observables, themselves rest on theoretical claims that are without justification. Still, a philosopher might propose a sort of inductive justification of theory-dependent scientific inductions. Let the inductive procedures of science be as theory-dependent as you like, and let the justifications offered for individual inferences by scientists be as faulty as the empiricist claims. The fact remains that the (theory-dependent) methodology of science gives evidence of being instrumentally reliable. Let *that* constitute the justification for the inferences which scientists make. The thesis that the methodology of science is instrumentally reliable is, after all, a thesis about observable phenomena. It is, moreover, well confirmed by the observational evidence presented by the recent history of science and technology. Since no abductive inference objectionable from an empiricist perspective is required to establish the generalization that scientific methodology is instrumentally reliable, we may accept this generalization and then apply it to justify the acceptance of the inductive generalizations scientists arrive at by employing the scientific method. Even though the theoretical reasoning that underlies inductive inferences about observables may not be justificatory, a second-order induction about the instrumental reliability of such reasoning might still afford a justification for that part of scientific practice that is supposed to be immune from the empiricist's selective skepticism.

It is very doubtful that this application of the inductive justification of induction can help the empiricist we are considering to avoid the conclusion that inductive generalizations in science about observables are unjustified. The hypothesis that scientific methodology is instrumentally reliable (henceforth, the "reliability hypothesis") is itself an inductive generalization about observable phenomena. If, as I have suggested earlier, the confirmation or disconfirmation of such generalizations typically presupposes theoretical considerations of the sort our empiricist cannot accept, then we should expect that this might be true of the confirmation of the reliability hypothesis itself. If this is so, then the effort to circumvent the empiricist's conclusion that inductive generalizations in science are unjustified because they are theory-dependent, by appealing to the confirmation of the reliability hypothesis, will have failed. The reliability hypothesis will itself be unjustified by the standards of the empiricist we are considering.

I earlier suggested that theory-dependent considerations enter into the confirmation or disconfirmation of inductive generalizations in science in two related ways. In the first, theoretical considerations are decisive in solving what Goodman calls the problem of "projectability." Given any finite body of observational data, there are infinitely many different generalizations about observables that are logically compatible with them. Theoretical considerations dictate the choice of a relatively small, finite number of these generalizations as "projectable," that is, as worthy of serious scientific and experimental consideration. Moreover, when the experimental evidence for or against such projectively appropriate generalizations is assessed, theoretical considerations are crucial in determining the degree of confirmation or disconfirmation which those generalizations receive, given any particular body of observational evidence. If this is so, then we might expect to be able to discern the effects of both sorts of theory-dependent judgments in the special case of the confirmation of the reliability hypothesis.

Consider first the issue of the degree of confirmation of the reliability hypothesis. The hypothesis that the scientific method is instrumentally reliable asserts that that method tends to produce acceptance of instrumentally reliable theories. The reliability of a theory in turn is a matter not only of its past predictive successes but also of its future predictive success. Now the observational evidence that supports the reliability hypothesis consists of the past and present predictive successes of (many of) the theories whose acceptance has been dictated by the scientific method. For these past successes to count as evidence for the instrumental reliability of the scientific method, they surely must be understood first as counting as evidence for the future (approximate) instrumental reliability of most of the theories in question.

The conviction that the methods of science are instrumentally reliable turns on the belief that those methods have led us to accept theories that tended themselves to be instrumentally reliable. We can make this latter judgment only if we take the past predictive successes of the relevant theories as evidence for their future instrumental reliability; that is, only if we are already prepared to make the ordinary scientific judgment that past predictive successes of the sort actually available warrant our belief in the inductive generalizations about observables embodied in the theories in question. But this is just the sort of theory-dependent judgment which the reliability hypothesis is supposed to justify. If the ordinary scientific justifications for assigning the generalizations in question a high degree of confirmation are inadequate because they depend upon abductions to theoretical explanations, then the second-order inductive justification of scientists' inductions by appeal to the reliability hypothesis fails to help. The decision to assign the reliability hypothesis a high degree of confirmation on the available evidence rests upon the very theory-dependent judgments about the degree of confirmation of ordinary scientific theories which the empiricist we are considering cannot accept as justificatory.

We may also see how theoretical considerations regarding "projectability" are involved in the confirmation of the reliability hypothesis. When philosophers of whatever persuasion assert that the methods of science are instrumentally (or theoretically, for that matter) reliable, their claim is of very little interest if nothing can be said about which methods are the methods in question. Indeed, without at least a preliminary specification of the methods in question, it would be difficult to have any evidence whatsoever for the reliability thesis. Moreover, it will not do to countenance as 'methods of science' just any regularities that may be discerned in the practice of scientists. If the reliability thesis is to be correctly formulated, one must identify those features of scientific practice that contribute to its instrumental reliability. This is a nontrivial intellectual problem, as one may see by examining the various different attempts—behaviorist, reductionist, and functionalist—to explain what a *scientific* foundation for psychology would look like.

Insofar as the confirmation of the reliability hypothesis is concerned, the issue is not so much a matter of how easy or difficult it is to identify the reliability-making features of scientific practice, but rather over what sorts of considerations would have to go into a justification for a proposed identification of those features. Recall that we are considering the options open to the empiricist who rejects abductive inferences as nonjustificatory but who agrees that the actual inductive methods of science (the instrumentally reliable methods) are theory-dependent and rest in practice upon abductive inferences. It is reasonable to ask of this empiricist—as it would be reasonable to ask of any other philosopher who had identified the same theory-dependent methods as the methods of sciences—what justification can be offered for the identification of these particular methods as the reliability-making features of scientific practice.

The problem of providing a justification for a particular proposed identification of such features represents, as regards the formulation of the reliability hypothesis, a special case of the problem of projectability. This may be seen easily if we employ a variant of the empiricists' favorite argument that theory choice is underdetermined by observational data.

Suppose that you believe that past scientific practice has certain reliability-making general features which should form the basis for a suitable
formulation of the reliability hypothesis. There have been, to date, only
finitely many methodological judgments in the whole history of science.
Even if you know which of these judgments have contributed to the reliability of past scientific practice, there will still be infinitely many different methodologies—infinitely many different sets of principles for theory
choice, experimental design, data assessment, and so forth—which
would have dictated the conclusions of those finitely many past methodological judgments. The choice of any one of these infinitely many
methodologies represents a particular solution to the problem of projectability for the investigator interested in finding an appropriate formulation of the reliability hypothesis. Alternative choices yield different
versions of the reliability hypothesis and represent different estimates of
what the reliability-making *general* features of past scientific practice
have been.

   If what I have suggested earlier is true, then the solution to this particular case, of the problem of projectability, might be expected to depend
upon *theoretical* considerations. Indeed, this proves to be the case. Remember that the empiricist we are considering accepts the ordinary
theory-dependent methods of the working scientist as the reliability-
making features of scientific practice. Let us consider an illustrative example of such methods. It is by now widely acknowledged that sound scientific methodology dictates that "measurement procedures" for physical
magnitudes should be revised in light of new theoretical "discoveries." (I
use quotation marks to indicate that the empiricist need not take the
notions of measurement or theoretical discovery at face value. What is
important is that the application of this principle in practice has a significant effect upon the inductive generalizations about observables which
scientists accept.) Let P be the methodological principle that says that one
should follow the dictates of the best confirmed theory in (re)designing
measurement procedures. What justifies us in taking P to be one of the
reliability-making features of scientific practice? Why should we not sub-
sume the finitely many cases to date of successful applications of this
principle under some other quite different maxim with which they are all
consistent?

   Recalling that an appeal to the reliability hypothesis is inappropriate
here, since what is at issue is the formulation and confirmation of that
hypothesis, it is hard to see how our reasons for accepting P as reliability-
making could be other than a summary of the ordinary reasons scientists
have for accepting various applications of P. But these are theory-dependent reasons—roughly, they amount to the idea that the best theories

represent results of the best (abductive) inferences regarding the unobservable magnitudes in question, and that, therefore, these theories are likely to provide approximately true accounts of how to measure those magnitudes. But, theoretical reasons of this sort are just those which the empiricist considers nonjustificatory. Worse yet, if we are to accept *P*, and not just some particular applications of *P*, as reliability-making it would seem that our justification for accepting *P* must involve not just the scientists' theoretical reasons for particular applications of *P* but the scientific realist's reasons for thinking *P* generally reliable. If the empiricist forgoes appeals to the abductive inferences of ordinary scientific practice on the grounds that such inferences are nonjustificatory, then it is hard to see how she can make scientifically sound judgments about which methods are scientific or about how even to formulate the reliability hypothesis.

It is worth noting that the empiricist we are considering gets into this particular difficulty largely because she accepts the results of recent philosophical and historical scholarship, which strongly suggest that the real methods of science are theory-dependent and rest in practice on abductive inferences of the sort unacceptable to empiricists. What appears to be true is that the consistent empiricist cannot hold both that the inductive methods of scientists are justified, insofar as generalizations about observables are concerned and, at the same time, accept the best recent work on the question of what those methods actually are.

I conclude that the empiricist who rejects abductive inferences is probably unable to avoid, in any plausible way, the conclusion that the inductive inferences that scientists make about observables are unjustified. Nevertheless, even if this is so, Fine's criticism of abductive arguments for realism still has force. If what is at issue is the legitimacy of abductive inferences to theoretical explanations in general, then there is a kind of circularity in the appeal to a particular abduction of this sort in the defense of scientific realism. I suggested earlier in this paper that while standard rebuttals to empiricist antirealism provide some reason to believe that scientific realism is true, these rebuttals fail to respond to the strong epistemological challenge that empiricist antirealism offers. Should we take the circularity which Fine discerns to indicate that the same is true for the abductive argument for scientific realism as a component in the best explanation for the instrumental reliability of scientific method? I want to argue that the answer should be no.

If abduction were prima facie suspect, in the way that palm reading or horoscope casting now are, then surely it would be inappropriate to appeal to some particular abductive inference in defense of abductive inference in general. Abduction is, however, prima facie legitimate; it is seen as suspect only in the light of certain distinctly empiricist epistemo-logical considerations. To assess the import of the circularity of appealing to abduction in replying to empiricist antirealism, we must examine more closely the relation between the particular abductive inferences in question, and the empiricist's arguments against realism.

I suggest that an assessment of the import of the circularity in question should focus not on the legitimacy of the realist's abductive inference considered in isolation, but rather on the relative merits of the overall accounts of scientific knowledge which the empiricist and the realist defend. Such an assessment strategy is familiar from many areas of intellectual inquiry, scientific and scholarly. Defenders of rival positions often reach their distinctive conclusions via forms of inference which their rivals think unjustified. The pair-wise theory-neutral procedure for addressing such disputes typically consists in an assessment of the overall adequacy of the theories put forward, rather than in an assessment of the particular controversial inference forms considered in isolation.

If we consider the present dispute in this light, then there are two especially important considerations. First, the empiricist's objection to abductive inferences (at least to those that yield conclusions about unobservable phenomena) rests upon the powerful and sophisticated epistemological argument rehearsed in my discussion of empiricist antirealism. That argument depends upon the evidential indistinguishability thesis. Moreover, the evidential indistinguishability thesis itself is put forward by empiricists (tacitly or explicitly) on the understanding that it captures the truth reflected in the doctrine of knowledge empiricism: the doctrine that all factual knowledge must be grounded in observation. If either knowledge empiricism is basically false, or the indistinguishability thesis represents a seriously misleading interpretation of it, then the empiricist's argument against abduction to theoretical explanation fails.

Second, the empiricist aims at a selectively skeptical account of scientific knowledge: knowledge of unobservables is impossible, but inductive generalizations about observables are sometimes epistemologically legitimate. It turns out, however, that the empiricist's commitment to knowledge empiricism, together with her adoption of the evidential indistinguishability thesis as an interpretation of it, threaten to dictate the unwelcome and implausible conclusion that even inductive inferences regarding observables are always unjustified.

The rebuttals to empiricist antirealism discussed earlier strengthen the case for realism as an account of the structure of scientific knowledge, yet they provide no direct argument either against knowledge empiricism or against the evidential indistinguishability thesis as an interpretation of it. The situation of the abductive argument for scientific realism sketched previously is quite different. If we accept the abductive inference to a distinctly realistic account of scientific methodology, then we can see *why* the evidential indistinguishability thesis is false. Moreover, we can see that the distinctly realistic conception of scientific methodology retains the central core of the doctrine of knowledge empiricism: all factual knowledge *does* depend upon observation; there are no a priori factual statements immune from empirical refutation.

I think it is fair to say that, given the difficulties that plague empiricist antirealism in the philosophy of science, the only philosophically cogent reason for rejecting scientific realism in favor of instrumentalism, or some other variant of empiricism, lies in the conviction that only from an empiricist perspective can one be faithful to the basic idea that factual knowledge must be experimental knowledge, that is, to the grain of truth in knowledge empiricism. The abductive argument for scientific realism that we are considering is best thought of as a component of an alternative realistic conception of scientific knowledge which preserves the empiricist insight that factual knowledge rests on the senses without the cost of an inadequate and potentially wholly skeptical treatment of scientific inquiry.

I suggested (in the section "Scientific Realism and Metaphilosophy") that the crucial feature of this alternative conception of knowledge is its naturalism. In particular, the special relation of the senses to knowledge is seen, in this conception, as resting on logically contingent facts about the role of the senses in the reliable production or regulation of belief. Here, an analogy between the naturalistic defense of scientific realism against empiricist antirealism and the naturalistic defense of knowledge of external objects against empiricist phenomenalism is revealing. The phenomenalist rejects realism about (observable) external objects, relying on an application of the sense-datum version of the evidential indistinguishability thesis. The indistinguishability thesis itself is understood as the appropriate interpretation of the fundamental truth embodied in the doctrine of knowledge empiricism. The causal theorist of knowledge does not reject the basic doctrine of the epistemic primacy of the senses but, instead, insists that the truth of that doctrine, insofar as it concerns perceptual knowledge, is really a reflection of the logically contingent fact that the senses are causally reliable detectors of external objects. Sensory experience provides reliable evidence for propositions only when it arises from suitable causal connections to the subject matter of the propositions in question. The sense-datum form of the indistinguishability thesis is, therefore, false and inadequately expresses the fundamental truth of knowledge empiricism.

The causal theorist's critique of phenomenalism rests upon what her

empiricist opponent would characterize as an illegitimate abductive inference to external objects as the explanation for facts about sensations. The
causal theorist's position does not, however, stand or fall on the strength
of that abduction taken in isolation. Instead, the alternative empiricist
and naturalist conceptions of knowledge and, especially, of the epistemic
role of the senses, must be evaluated as rival philosophical theories. The
very grave difficulties that phenomenalism faces in explaining ordinary
perceptual knowledge strongly suggest that the naturalist's causal theory
of perceptual knowledge is preferable.

The situation with respect to the dispute between the empiricist antirealist and the scientific realist who subscribes to the argument sketched
earlier (in section "Defending Scientific Realism") is exactly analogous.
The antirealist's position rests upon an application of the indistinguishability thesis, which in turn is offered as an explication of knowledge
empiricism. The scientific realist, like the causal theorist of perception,
accepts the insight of knowledge empiricism while denying that the indistinguishability thesis captures that insight. The causal theorist maintains
that the truth of knowledge empiricism, insofar as it applies to perceptual
knowledge, is a reflection of a logically contingent fact about the reliability of the senses as detectors. Analogously, the scientific realist maintains
that the truth of knowledge empiricism, insofar as experimental knowledge in the sciences is concerned, is a reflection not only of the logically
contingent reliability of the senses as detectors but also of the logically
and historically contingent emergence of a theoretical tradition relevantly
approximately true enough to make theory-dependent experimental practice a reliable mechanism for belief regulation. Like the causal theorist's
rebuttal to phenomenalism, the scientific realist's rebuttal to empiricist
antirealism rests upon what her opponent would regard as an illegitimate
abductive inference. In this case, like the previous one, however, the scientific realist's position does not stand or fall on the strength of that
abduction considered in isolation. Rather, what is to be assessed are the
relative merits of empiricist epistemology and the emerging naturalistic
epistemology of which the realist's conception of scientific knowledge is
one of the more distinctive and controversial parts.

In this regard, it is worth remarking that the plausibility of knowledge
empiricism has, no doubt, always rested upon two considerations: a recognition of the central causal role of the senses in information gathering,
and a recognition of the success of experimental science. It is doubtful if
consistent empiricism can recognize either of these phenomena. If this
proves to be the case, then the alternative realistic and naturalistic conception of the epistemic role of the senses must surely capture what truth
there is in knowledge empiricism.

THE BEARING OF THE HISTORY OF SCIENCE UPON ARGUMENTS FOR REALISM

Let us turn now to the question of how evidence from the history of science bears upon the arguments for scientific realism that we have been discussing. I have emphasized the important role that, according to the version of naturalistic and realistic epistemology discussed in this paper, was played by the historically contingent emergence of research traditions embodying suitably approximately true theories of unobservables. If I am right, it is to the successive development of the approximate truths (theoretical as well as instrumental) embodied in these traditions that we owe the instrumental reliability of current scientific practice. Although it is no part of my thesis that this development was progressive in all particular instances, or uniform with respect to different disciplines, subdisciplines, or even problem areas within subdisciplines, it is essential to the thesis I am defending that there be some measure of referential continuity and successive approximation to the truth in the history of recent science. I emphasized earlier in the present essay that if continuity of reference and methodology could not be established in many cases in the history of modern science, the sort of realism I am defending would be strongly undermined.

Because considerations of historical continuity are central to the abductive argument for scientific realism we are considering here, I think it important to indicate ways in which historical continuity is *not* involved in that argument. In the first place, it is not a consequence of the position advocated here on behalf of the realist that a successful pattern of inductive generalization at the observational level must *always* rest upon the acceptance of relevantly approximately true background theories. For any inductive enterprise to be successful, there must be an appropriate correspondence between the categories, in terms of which phenomena are classified, and their relevant causal powers. There is, however, nothing to prevent scientists or others from hitting upon categories appropriate to some limited class of generalizations by chance rather than through theoretical understanding.

In mature sciences, however, scientists do not solve the problem of projectability by the specification of some relatively fixed sets of projectable properties or predicates, whether theoretical or observational. Instead, we possess a *methodology* for exploiting the full descriptive resources of our *theoretical* concepts, to guide inductive inferences at the observational level. Instead of assessing the projectability of particular predicates, we are able to assess the projectability of theoretically characterizable patterns in observational data: we count as projectable any pattern in observational data that corresponds to a theoretical hypothesis that is plausible in the light of the current total science. Moreover, we take such a hypothesis to represent the inductive generalization about observables that corresponds to the observational consequences derivable from the hypothesis itself, *together with* the theories that constitute the existing total science. Once such a hypothesis has been accepted, we countenance further expansion and modification of the inductive generalizations about observables that it warrants as our total science itself changes and develops.[35] Thus we are able to identify as projectable an extraordinary variety of patterns among observables representing empirical generalizations of great power, scope, and precision.

In addition to the methods for identifying inductively appropriate empirical generalizations, the methods employed in mature sciences for the experimental and observational testing of such generalizations—methods for the design of experiments and of instrumentation, for the establishment of appropriate controls, and for the assessment of degrees of confirmation—are also profoundly theory-dependent. It is the instrumental reliability of all of these various theory-dependent methods—methods whose characteristic reliability is displayed typically only in mature (and, often, relatively recent) science—for which, according to the argument we are considering, the only plausible explanation rests upon a realistic conception of scientific knowledge. What is claimed is that when, in the historical development of any particular science, its theory-dependent methodological practices come to display the sort of intricacy *and* instrumental reliability characteristic, say, of modern physical or chemical practice, only the realistic account of scientific knowledge described earlier in this essay will provide an adequate explanation of that reliability. No claim is made that the more limited inductive success of earlier scientific practice must always be explained in the same way. Nor is it claimed, even in the case of inquiry in mature sciences, that the approximate theoretical knowledge upon which the instrumental reliability of methodology depends must represent *fundamental* knowledge, or knowledge of the *ultimate* essences of the phenomena in question. The abductive argument for realism does not require that the approximate theoretical knowledge which scientists possess must embody correct answers to those questions that scientists or philosophers might consider most basic or fundamental. All that is claimed is that the instrumental reliability of the methodology of mature sciences depends upon the development of a theoretical tradition that embodies approximate knowledge of unobservable as well as observable phenomena. It is this claim, after all, which the empiricist denies.

Similarly, it is *not* a thesis of the version of scientific realism defended here that there is one completely true theory which would be the "asymptotic limit" of scientific theorizing if science were pursued long enough. As antireductionist materialists have long insisted, there is no reason to believe that true theories are all special cases of some most fundamental theory, even if materialism is true. Different levels of description or of functional organization characterize different, perfectly real, natural phenomena. Even if one understands by a "theory" something like a "total science"—a set of sentences that may embody descriptions of phenomena at various levels of functional or structural organization—it does not follow from the sort of realism defended here that *the* true theory would be the asymptotic limit of scientific inquiry.

In the first place, even for theories that describe phenomena at the same level of organization, and even for theories that are in some sense complete in their description of the relevant phenomena, it does not follow from a realistic conception of science that there must be a single true theory. In particular, it does not follow that there must be a single true ontology for the most basic level of physical theory (assuming that there is such a level). What is entailed is that if there are two entirely true and suitably complete theories of basic physical phenomena, they must be ontologically equivalent in the sense that the entities, powers, properties, states, and so forth, which form the ontology of any one, must themselves be causally realized by the entities that form the ontology of the other. On the standard positivist analysis of ontological equivalence, this would entail that the two theories must be syntactically reducible to each other and, thus, that they be linguistic variants of the *same* theory. Such a positivist analysis of ontological equivalence is mistaken and is, in fact, simply a reflection of an antirealist conception of causal relations. On a realist conception of ontological equivalence, no such conclusion follows, so that it is perfectly conceivable that scientific research might converge to one of two such theories, while the ontological conceptions central to the second might be quite literally inexpressible, given the descriptive resources of the first theoretical tradition.

In the second place, it is no part of the realistic conception of science defended here that any such convergence to (even one version of) the exact truth need occur even in the ideal limit of actual scientific practice. There are any number of ways in which our understanding might be forever "bounded away" from the exact truth about some (or even all) aspects of nature.[36]

Finally, the evidential connection is quite subtle between, on the one hand, the historical evidence for continuity of theoretical semantics and for continuity of methods in mature sciences, and, on the other, the thesis of scientific realism. Because scientific realists hold that progress in mature sciences is a reflection of theoretical as well as instrumental progress and, indeed, that instrumental progress often depends upon theoretical progress, it is essential to the empirical case for realism that historical evidence support the claim that there is the relevant sort of semantic and methodological continuity in the history of mature sciences. For example, it must be possible to see greater continuity and commensurability across "scientific revolutions" than Kuhn acknowledges. When the history of science provides evidence of semantic or methodological continuity in mature sciences, the realist will typically hold that a realist conception of scientific knowledge—together with the appropriate sort of referential continuity for theoretical terms—provides the best explanation for the historical evidence in question.[37]

But it is *not* part of the strategy for the defense of realism described here to suggest that any substantial prima facie evidence *for* scientific realism is provided merely by consideration of historical evidence of this sort. The two chief rivals of scientific realism, empiricism and constructivism, are each capable of providing plausible explanations for the apparent semantic and methodological continuity in the history of well-developed and mature sciences. Indeed, they offer variations on the same explanation: the continuity in question is a manifestation of linguistic, conceptual, and methodological *conventions* (see the earlier section on the similarities between empiricism and constructivism). If we focus our attention solely on the historical evidence for semantic and methodological continuity in the history of science, there seems little reason to prefer the realist's explanation to that of the constructivist or the empiricist.

According to the realist position discussed here, the choice between the competing explanations for apparent semantic and methodological continuity in mature sciences must rest upon other considerations. Neither the empiricist nor the constructivist can explain the most striking feature of the recent history of science, that is, the instrumental reliability of its methods. Only scientific realism provides the resources for explaining this crucial historical phenomenon. It is for this reason that realism is to be preferred to rival accounts of scientific knowledge, and for this reason that the realist account of semantic and methodological continuity is to be preferred to the alternative account presented in various forms by empiricists and constructivists.

The positive evidence for scientific realism thus rests primarily on features of scientific practice that would be discernible even if one limited one's examination to very recent science. According to the realist, realism provides the only acceptable explanation for the current instrumental reliability of scientific methodology in mature sciences. Realism does, however, entail interesting conclusions about historical development within mature sciences—that is, within those sciences in which theoretical considerations contribute significantly to a high level of instrumental reliability of method. For many sciences, especially the physical sciences, the period of maturity, in this sense, begins long before the recent past. Thus, historical studies of such sciences, of, for example, the extent of semantic and methodological continuity in the history of those sciences, are evidentially relevant to the issue of realism. Insofar as a realist perspective proves fruitful in understanding the history of mature sciences, that would provide further evidence for realism; the primary role of historical studies in this area, however, is to subject the claims of realists to possible disconfirmation by historical evidence rather than to provide new kinds of positive evidence favoring realism over its rivals.

There is one important respect in which consideration of the implications of scientific realism regarding the nonrecent history of science does provide additional justification for the acceptance of realism, but here the connection with the assessment of the historical evidence for realism is indirect. What I have in mind is this: it is by reflection on the historical implications of a realist conception of scientific knowledge that we are able to see (a) that the reliability (instrumental or theoretical) of the scientific method rests upon the logically and historically contingent emergence of a suitably approximately true tradition, and (b) that judgments of the plausibility of theories relative to such a tradition are evidential. It is these doctrines, in turn, which enable us to see that the evidential indistinguishability thesis is false, that a theory-dependent methodology need not be merely a construction procedure, and that a realist conception of the epistemology of science can be integrated into, and can serve to justify, a broader naturalistic conception of epistemology and of philosophy itself. It is upon these last two considerations that the case for scientific realism ultimately rests, and it is in its contribution to a naturalistic conception of philosophy that scientific realism makes its greatest contribution to an understanding of the nature of knowledge.

NOTES

1. B. van Fraassen, *The Scientific Image* (Oxford: The Clarendon Press, 1980).

2. T. Kuhn, *The Structure of Scientific Revolutions* (Chicago: University of Chicago Press, 1970).

3. N. R. Hanson, *Patterns of Discovery* (Cambridge: Cambridge University Press, 1958).

4. J. J. C. Smart, *Philosophy and Scientific Realism* (London: Routledge and Kegan Paul, 1963).

5. Kuhn, *The Structure of Scientific Revolutions*.

6. J. Bennett, *Locke, Berkeley, Hume* (Oxford: Oxford University Press, 1971).

7. See, for example, G. Maxwell, "The Ontological Status of Theoretical Entities," in *Scientific Explanation, Space and Time*, ed. H. Feigl and G. Maxwell (Minneapolis: University of Minnesota Press, 1963).

8. G. Harman, "The Inference to the Best Explanation," *Philosophical Review* (January 1965).

9. This may be the argument that Putnam, *Meaning and the Moral Sciences* (London: Routledge and Kegan Paul, 1978), attributes to Boyd, *Realism and Scientific Epistemology* (Cambridge: Cambridge University Press, forthcoming).

10. R. Boyd, "Scientific Realism and Naturalistic Epistemology," *PSA* (1980), vol. 2, ed. P. D. Asquith and R. N. Giere.

11. R. Boyd, "Metaphor and Theory Change," *Metaphor and Thought*, ed. Andrew Ortony (Cambridge: Cambridge University Press, 1979); "Scientific Realism and Naturalistic Epistemology"; *Realism and Scientific Epistemology*.

12. This is Kuhn's term; see Kuhn, *The Structure of Scientific Revolutions*.

13. Ibid.

14. Boyd, "Metaphor and Theory Change."

15. Kuhn, *The Structure of Scientific Revolutions*, 101-102; see Boyd, "Metaphor and Theory Change," for a discussion.

16. R. Carnap, *Meaning and Necessity* (Chicago: University of Chicago Press, 1950).

17. See M. Schlick, "Positivism and Realism," *Erkenntnis* 3 (1932-33), in *Logical Positivism*, ed. A. J. Ayer (New York: Free Press, 1959), for an anticipation of Carnap's later position.

18. Boyd, "Metaphor and Theory Change," especially 397-398.

19. N. Goodman, *Fact, Fiction and Forecast*, 3d ed. (Indianapolis and New York: Bobbs-Merrill Co., 1973).

20. For further discussion of this issue, see W. V. O. Quine, "Natural Kinds," in W. V. O. Quine, *Ontological, Relativity and Other Essays* (New York: Columbia University Press, 1969); and Boyd, "Metaphor and Theory Change"; R. Boyd, "Materialism without Reductionism: What Physicalism Does Not Entail," in *Readings in Philosophy of Psychology*, vol. 1, ed. Ned Block (Cambridge: Harvard University Press, 1980); and Boyd, "Scientific Realism and Naturalistic Epistemology."

21. See Boyd, "Scientific Realism and Naturalistic Epistemology," and Boyd, *Realism and Scientific Epistemology*, for discussion.

22. R. Boyd, "Determinism, Laws and Predictability in Principle," *Philosophy of Science* 39 (1972); "Realism, Underdetermination and A Causal Theory of Evidence," *Nous* (March 1973); "Metaphor and Theory Change"; "Scientific Realism and Naturalistic Epistemology"; "Materialism without Reductionism: Non-Humean Causation and the Evidence for Physicalism," in *The Physical Basis of Mind*, ed. R. Boyd (Cambridge, Mass.: Harvard University Press, forthcoming); *Realism and Scientific Epistemology* (Cambridge: Cambridge University Press, forthcoming).

23. Boyd, "Scientific Realism and Naturalistic Epistemology."

24. See Boyd, "Metaphor and Theory Change"; "Materialism without Reduc-

tionism: What Physicalism Does Not Entail"; "Scientific Realism and Naturalistic Epistemology"; "Materialism without Reductionism: Non-Humean Causation"; and *Realism and Scientific Epistemology,* for discussion of these points.

25. Boyd, "Scientific Realism and Naturalistic Epistemology"; see also Quine, "Natural Kinds."

26. H. Feigl, "Some Major Issues and Developments in the Philosophy of Science of Logical Empiricism," in *Minnesota Studies in the Philosophy of Science,* vol. 1, ed. H. Feigl and M. Scriven (Minneapolis: University of Minnesota Press, 1956); S. Kripke, "Naming the Necessity," in *The Semantics of Natural Language,* eds. G. Harman and D. Davidson (Dordrecht: D. Reidel, 1972); H. Putnam, "The Meaning of 'Meaning,'" in *Mind, Language and Reality,* ed. H. Putnam, Philosophical Papers, vol. 2 (Cambridge: Cambridge University Press, 1975).

27. Boyd, "Metaphor and Theory Change"; "Scientific Realism and Naturalistic Epistemology"; *Realism and Scientific Epistemology.*

28. Boyd, "Metaphor and Theory Change"; "Scientific Realism and Naturalistic Epistemology."

29. Boyd, "Metaphor and Theory Change."

30. H. Field, "Theory Change and the Indeterminacy of Reference," *Journal of Philosophy* 70 (1973).

31. E. Nagel, *The Structure of Science* (New York: Harcourt Brace, 1961), 151-152.

32. For a discussion of limitations of this process of successive approximation see Boyd, "Scientific Realism and Naturalistic Epistemology."

33. See ibid., for a discussion of the relation between scientific realism and other recent naturalistic trends in epistemology.

34. See "The Natural Ontological Attitude" in this volume. I am extremely grateful to Professor Fine for the opportunity to read a prepublication copy of this paper.

35. For a more precise discussion, see Boyd, "Scientific Realism and Naturalistic Epistemology."

36. Ibid.

37. See Boyd, "Metaphor and Theory Change," for a more carefully qualified formulation of this claim.

4

The Natural Ontological Attitude

Arthur Fine

> Let us fix our attention out of ourselves as much as possible;
> let us chace our imagination to the heavens, or to the utmost
> limits of the universe; we never really advance a step beyond
> ourselves, nor can conceive any kind of existence, but those
> perceptions, which have appear'd in that narrow compass.
> This is the universe of the imagination, nor have we any idea
> but what is there produced.
> —Hume, *Treatise*, Book 1, Part II, Section VI

Realism is dead. Its death was announced by the neopositivists who realized that they could accept all the results of science, including all the
members of the scientific zoo, and still declare that the questions raised
by the existence claims of realism were mere pseudoquestions. Its death
was hastened by the debates over the interpretation of quantum theory,
where Bohr's nonrealist philosophy was seen to win out over Einstein's
passionate realism. Its death was certified, finally, as the last two generations of physical scientists turned their backs on realism and have managed, nevertheless, to do science successfully without it. To be sure,
some recent philosophical literature, and some of the best of it represented by contributors to this book, has appeared to pump up the ghostly
shell and to give it new life. But I think these efforts will eventually be
seen and understood as the first stage in the process of mourning, the
stage of denial. This volume contains some further expressions of this

denial. But I think we shall pass through this first stage and into that of acceptance, for realism is well and truly dead, and we have work to get on with, in identifying a suitable successor. To aid that work I want to do three things in this essay. First, I want to show that the arguments in favor of realism are not sound, and that they provide no rational support for belief in realism. Then, I want to recount the essential role of nonrealist attitudes for the development of science in this century, and thereby (I hope) to loosen the grip of the idea that only realism provides a progressive philosophy of science. Finally, I want to sketch out what seems to me a viable nonrealist position, one that is slowly gathering support and that seems a decent philosophy for postrealist times.[1]

ARGUMENTS FOR REALISM

Recent philosophical argument in support of realism tries to move from the success of the scientific enterprise to the necessity for a realist account of its practice. As I see it, the arguments here fall on two distinct levels. On the ground level, as it were, one attends to particular successes; such as novel, confirmed predictions, striking unifications of disparate-seeming phenomena (or fields), successful piggybacking from one theoretical model to another, and the like. Then, we are challenged to account for such success, and told that the best and, it is slyly suggested, perhaps, the *only* way of doing so is on a realist basis. I do not find the details of these ground-level arguments at all convincing. Larry Laudan has provided a forceful and detailed analysis which shows that not even with a lot of hand waving (to shield the gaps in the argument) and charity (to excuse them) can realism itself be used to explain the very successes to which it invites our attention.[2] But there is a second level of realist argument, the methodological level, that derives from Popper's attack on instrumentalism as inadequate to account for the details of his own, falsificationist methodology. Arguments on this methodological level have been skillfully developed by Richard Boyd,[3] and by one of the earlier Hilary Putnams.[4] These arguments focus on the methods embedded in scientific practice, methods teased out in ways that seem to me accurate and perceptive about ongoing science. We are then challenged to account for why these methods lead to scientific success and told that the best, and (again) perhaps, the only truly adequate way of explaining the matter is on the basis of realism.

I want to examine some of these methodological arguments in detail to display the flaws that seem to be inherent in them. But first I want to point out a deep and, I think, insurmountable problem with this entire strategy of defending realism, as I have laid it out above. To set up the problem, let me review the debates in the early part of this century over the foundations of mathematics, the debates that followed G. Cantor's introduction of set theory. There were two central worries here, one over the meaningfulness of Cantor's hierarchy of sets insofar as it outstripped the number-theoretic content required by L. Kronecker (and others); the second worry, certainly deriving in good part from the first, was for the consistency (or not) of the whole business. In this context, D. Hilbert devised a quite brilliant program to try to show the consistency of a mathematical theory by using only the most stringent and secure means. In particular, if one were concerned over the consistency of set theory, then clearly a set-theoretic proof of consistency would be of no avail. For if set theory were inconsistent, then such a consistency proof would be both possible and of no significance. Thus, Hilbert suggested that finite constructivist means, satisfactory even to Kronecker (or L. Brouwer) ought to be employed in metamathematics. Of course, Hilbert's program was brought to an end in 1931, when K. Gödel showed the impossibility of such a stringent consistency proof. But Hilbert's idea was, I think, correct even though it proved to be unworkable. Metatheoretic arguments must satisfy more stringent requirements than those placed on the arguments used by the theory in question, for otherwise the significance of reasoning about the theory is simply moot. I think this maxim applies with particular force to the discussion of realism.

Those suspicious of realism, from A. Osiander to H. Poincaré and P. Duhem to the 'constructive empiricism' of van Fraassen,[5] have been worried about the significance of the explanatory apparatus in scientific investigations. While they appreciate the systematization and coherence brought about by scientific explanation, they question whether acceptable explanations need to be true and, hence, whether the entities mentioned in explanatory principles need to exist.[6] Suppose they are right. Suppose, that is, that the usual explanation-inferring devices in scientific practice do not lead to principles that are reliably true (or nearly so), nor to entities whose existence (or near-existence) is reliable. In that case, the usual abductive methods that lead us to good explanations (even to 'the best explanation') cannot be counted on to yield results even approximately true. But the strategy that leads to realism, as I have indicated, is just such an ordinary sort of abductive inference. Hence, if the nonrealist were correct in his doubts, then such an inference to realism as the best explanation (or the like), while possible, would be of no significance— exactly as in the case of a consistency proof using the methods of an inconsistent system. It seems, then, that Hilbert's maxim applies to the debate over realism: to argue for realism one must employ methods more stringent than those in ordinary scientific practice. In particular, one must not beg the question as to the significance of explanatory hypotheses by assuming that they carry truth as well as explanatory efficacy.

There is a second way of seeing the same result. Notice that the issue over realism is precisely the issue as to whether we should believe in the reality of those individuals, properties, relations, processes, and so forth, used in well-supported explanatory hypotheses. Now what *is* the hypothesis of realism, as it arises as an explanation of scientific practice? It is just the hypothesis that our accepted scientific theories are approximately true, where "being approximately true" is taken to denote an extratheoretical relation between theories and the world. Thus, to address doubts over the reality of relations posited by explanatory hypotheses, the realist proceeds to introduce a further explanatory hypothesis (realism), itself positing such a relation (approximate truth). Surely anyone serious about the issue of realism, and with an open mind about it, would have to behave inconsistently if he were to accept the realist move as satisfactory.

Thus, both at the ground level and at the level of methodology, no support accrues to realism by showing that realism is a good hypothesis for explaining scientific practice. If we are open-minded about realism to begin with, then such a demonstration (even if successful) merely begs the question that we have left open ("need we take good explanatory hypotheses as true?"). Thus, Hilbert's maxim applies, and we must employ patterns of argument more stringent than the usual abductive ones. What might they be? Well, the obvious candidates are patterns of induction leading to empirical generalizations. But, to frame empirical generalizations, we must first have some observable connections between observables. For realism, this must connect theories with the world by way of approximate truth. But no such connections are observable and, hence, suitable as the basis for an inductive inference. I do not want to labor the points at issue here. They amount to the well-known idea that realism commits one to an unverifiable correspondence with the world. So far as I am aware, no recent defender of realism has tried to make a case based on a Hilbert strategy of using suitably stringent grounds and, given the problems over correspondence, it is probably just as well.

The strategy of arguments to realism as a good explanatory hypothesis, then, *cannot* (logically speaking) be effective for an open-minded nonbeliever. But what of the believer? Might he not, at least, show a kind of internal coherence about realism as an overriding philosophy of science, and should that not be of some solace, at least for the realist?[7] Recall, however, the analogue with consistency proofs for inconsistent systems. That sort of harmony should be of no solace to anyone. But for realism, I fear, the verdict is even harsher. For, so far as I can see, the arguments in question just do not work, and the reason for that has to do with the same question-begging procedures that I have already identified. Let me look closely at some methodological arguments in order to display the problems.

A typical realist argument on the methodological level deals with what I shall call the problem of the "small handful." It goes like this. At any time, in a given scientific area, only a small handful of alternative theories (or hypotheses) are in the field. Only such a small handful are seriously considered as competitors, or as possible successors to some theory requiring revision. Moreover, in general, this handful displays a sort of family resemblance in that none of these live options will be too far from the previously accepted theories in the field, each preserving the well-confirmed features of the earlier theories and deviating only in those aspects less confirmed. Why? Why does this narrowing down of our choices to such a small handful of cousins of our previously accepted theories work to produce good successor theories?

The realist answers this as follows. Suppose that the already existing theories are themselves approximately true descriptions of the domain under consideration. Then surely it is reasonable to restrict one's search for successor theories to those whose ontologies and laws resemble what we already have, especially where what we already have is well confirmed. And if these earlier theories were approximately true, then so will be such conservative successors. Hence, such successors will be good predictive instruments; that is, they will be successful in their own right.

The small-handful problem raises three distinct questions: (1) why only a small handful out of the (theoretically) infinite number of possibilities? (2) why the conservative family resemblance between members of the handful? and (3) why does the strategy of narrowing the choices in this way work so well? The realist response does not seem to address the first issue at all, for even if we restrict ourselves just to successor theories resembling their progenitors, as suggested, there would still, theoretically, always be more than a small handful of these. To answer the second question, as to why conserve the well-confirmed features of ontology and laws, the realist must suppose that such confirmation is a mark of an approximately correct ontology and approximately true laws. But how could the realist possibly justify such an assumption? Surely, there is no valid inference of the form "T is well-confirmed; therefore, there exist objects pretty much of the sort required by T and satisfying laws approximating to those of T." Any of the dramatic shifts of ontology in science will show the invalidity of this schema. For example, the loss of the ether from the turn-of-the-century electrodynamic theories demonstrates this at the level of ontology, and the dynamics of the Rutherford-Bohr atom vis-à-vis the classical energy principles for rotating systems demonstrates it at the level of laws. Of course, the realist might respond that there is no question of a strict inference between being well confirmed and being approximately true (in the relevant respects), but there is a probable inference of some sort. But of what sort? Certainly there is no probability relation that rests on inductive evidence here. For there is no independent evidence for the relation of approximate truth itself; at least, the realist has yet to produce any evidence that is independent of the argument under examination. But if the probabilities are not grounded inductively, then how else? Here, I think the realist may well try to fall back on his original strategy, and suggest that being approximately true provides the best explanation for being well confirmed. This move throws us back to the ground-level realist argument, the argument from specific success to an approximately true description of reality, which Lauden has criticized. I should point out, before looking at the third question, that if this last move is the one the realist wants to make, then his success at the method- ological level can be no better than his success at the ground level. If he fails there, he fails across the board.

The third question, and the one I think the realist puts most weight on, is why does the small-handful strategy work so well. The instrumentalist, for example, is thought to have no answer here. He must just note that it does work well, and be content with that. The realist, however, can ex- plain why it works by citing the transfer of approximate truth from predecessor theories to the successor theories. But what does this explain? At best, it explains why the successor theories cover the same ground as well as their predecessors, for the conservative strategy under considera- tion assures that. But note that here the instrumentalist can offer the same account: if we insist on preserving the well-confirmed components of ear- lier theories in later theories, then, of course the later ones will do well over the well-confirmed ground. The difficulty, however, is not here at all but rather is in how to account for the successes of the later theories in new ground or with respect to novel predictions, or in overcoming the anomalies of the earlier theories. And what can the realist possibly say in this area except that the theorist, in proposing a new theory, has hap- pened to make a good guess? For nothing in the approximate truth of the old theory can guarantee (or even make it likely) that modifying the theory in its less-confirmed parts will produce a progressive shift. The history of science shows well enough how such tinkering succeeds only now and again, and fails for the most part. This history of failures can scarcely be adduced to explain the occasional success. The idea that by extending what is approximately true one is likely to bring new approxi- mate truth is a chimera. It finds support neither in the logic of approximate truth nor in the history of science. The problem for the realist is how to explain the *occasional success* of a strategy that *usually fails.*[8] I think he has no special resources with which to do this. In particular, his usual fallback onto approximate truth provides nothing more than a gentle pillow. He may rest on it comfortably, but it does not really help to move his cause forward.

The problem of the small handful raises three challenges: why small, why narrowly related, and why does it work. The realist has no answer for the first of these, begs the question as to the truth of explanatory hypotheses on the second, and has no resources for addressing the third. For comparison, it may be useful to see how well his archenemy, the instrumentalist, fares on the same turf. The instrumentalist, I think, has a substantial basis for addressing the questions of smallness and narrowness, for he can point out that it is extremely difficult to come up with alternative theories that satisfy the many empirical constraints posed by the instrumental success of theories already in the field. Often it is hard enough to come up with even one such alternative. Moreover, the common apprenticeship of scientists working in the same area certainly has the effect of narrowing down the range of options by channeling thought into the commonly accepted categories. If we add to this the instrumentally justified rule, "If it has worked well in the past, try it again," then we get a rather good account, I think, of why there is usually only a small and narrow handful. As to why this strategy works to produce instrumentally successful science, we have already noted that for the most part it does not. Most of what this strategy produces are failures. It is a quirk of scientific memory that this fact gets obscured, much as do the memories of bad times during a holiday vacation when we recount all our "wonderful" vacation adventures to a friend. Those instrumentalists who incline to a general account of knowledge as a social construction can go further at this juncture, and lean on the sociology of science to explain how the scientific community "creates" its knowledge. I am content just to back off here and note that over the problem of the small handful, the instrumentalist scores at least two out of three, whereas the realist, left to his own devices, has struck out.[9]

I think the source of the realist's failure here is endemic to the methodological level, infecting all of his arguments in this domain. It resides, in the first instance, in his repeating the question-begging move from explanatory efficacy to the truth of the explanatory hypothesis. And in the second instance, it resides in his twofold mishandling of the concept of approximate truth: first, in his trying to project from some body of assumed approximate truths *to* some further and novel such truths, and second, in his needing genuine access to the relation of correspondence.

There are no general connections of this first sort, however, sanctioned by the logic of approximate truth, nor secondly, any such warranted access. However, the realist must pretend that there are, in order to claim explanatory power for his realism. We have seen those two agents infecting the realist way with the problem of the small handful. Let me show them at work in another methodological favorite of the realist, the "problem of conjunctions."

The problem of conjunctions is this. If T and T' are independently well-confirmed, explanatory theories, and if no shared term is ambiguous between the two, then we expect the conjunction of T and T' to be a reliable predictive instrument (provided, of course, that the theories are not mutually inconsistent). Why? challenges the realist, and he answers as follows. If we make the realist assumption that T and T', being well confirmed, are approximately true of the entities (etc.) to which they refer, and if the unambiguity requirement is taken realistically as requiring a domain of common reference, then the conjunction of the two theories will also be approximately true and, hence, it will produce reliable observational predictions. Q.E.D.

But notice our agents at work. First, the realist makes the question-begging move from explanations to their approximate truth, and then he mistreats approximate truth. For nothing in the logic of approximate truth sanctions the inference from "T is approximately true" and "T' is approximately true" to the conclusion that the conjunction "T · T'" is approximately true. Rather, in general, the tightness of an approximation dissipates as we pile on further approximations. If T is within ϵ, in its estimation of some parameter, and T' is also within ϵ, then the only general thing we can say is that the conjunction will be within 2ϵ of the parameter. Thus, the logic of approximate truth should lead us to the opposite conclusion here; that is, that the conjunction of two theories is, in general, *less* reliable than either (over their common domain). But this is neither what we expect nor what we find. Thus, it seems quite implausible that our actual expectations about the reliability of conjunctions rest on the realist's stock of approximate truths.

Of course, the realist could try to retrench here and pose an additional requirement of some sort of uniformity on the character of the approximations, as between T and T'.[10] It is difficult to see how the realist could do this successfully without making reference to the distance between the approximations and "the truth." For what kind of internalist requirement could possibly insure the narrowing of this distance? But the realist is in no position to impose such requirements, since neither he nor anyone else has the requisite access to "the truth." Thus, whatever uniformity-of-approximation condition the realist might impose, we could still demand to be shown that this leads closer to the truth, not farther away. The realist will have no demonstration, except to point out to us that it all works (sometimes!). But that was the original puzzle.[11] Actually, I think the puzzle is not very difficult. For surely, if we do not entangle ourselves with issues over approximation, there is no deep mystery as to why two compatible and successful theories lead us to expect their conjunction to be successful. For in forming the conjunction, we just add the reliable predictions of one onto the reliable predictions of the other, having antecedently ruled out the possibility of conflict.

There is more to be said about this topic. In particular, we need to address the question as to why we expect the logical gears of the two theories to mesh. However, I think that a discussion of the realist position here would only bring up the same methodological and logical problems that we have already uncovered at the center of the realist argument.

Indeed, this schema of knots in the realist argument applies across the board and vitiates every single argument at the methodological level. Thus my conclusion here is harsh, indeed. The methodological arguments for realism fail, even though, were they successful, they would still not support the case. For the general strategy they are supposed to implement is just not stringent enough to provide rational support for realism. In the next two sections, I will try to show that this situation is just as well, for realism has not always been a progressive factor in the development of science and, anyway, there is a position other than realism that is more attractive.

REALISM AND PROGRESS

If we examine the two twentieth-century giants among physical theories, relativity and the quantum theory, we find a living refutation of the realist's claim that only his view of science explains its progress, and we find some curious twists and contrasts over realism as well. The theories of relativity are almost singlehandedly the work of Albert Einstein. Einstein's early positivism and his methodological debt to Mach (and Hume) leap right out of the pages of the 1905 paper on special relativity.[12] The same positivist strain is evident in the 1916 general relativity paper as well, where Einstein (in Section 3 of that paper) tries to justify his requirement of general covariance by means of a suspicious-looking verificationist argument which, he says, "takes away from space and time the last remnants of physical objectivity."[13] A study of his tortured path to general relativity (see here the brilliant work of John Earman, following on earlier hints by Banesh Hoffmann)[14] shows the repeated use of this

Machist line, always used to deny that some concept has a real referent. Whatever other, competing strains there were in Einstein's philosophical orientation (and there certainly were others), it would be hard to deny the importance of this instrumentalist/positivist attitude in liberating Einstein from various realist commitments. Indeed, on another occasion, I would argue in detail that without the "freedom from reality" provided by his early reverence for Mach, a central tumbler necessary to unlock the secret of special relativity would never have fallen into place.[15] A few years after his work on general relativity, however, roughly around 1920, Einstein underwent a philosophical conversion, turning away from his positivist youth (he was forty-one in 1920) and becoming deeply committed to realism.[16] His subsequent battle with the quantum theory, for example, was fought much more over the issue of realism than it was over the issue of causality or determinism (as it is usually portrayed). In particular, following his conversion, Einstein wanted to claim genuine reality for the central theoretical entities of the general theory, the four-dimensional space-time manifold and associated tensor fields. This is a serious business for if we grant his claim, then not only do space and time cease to be real but so do virtually all of the usual dynamical quantities.[17] Thus motion, as we understand it, itself ceases to be real. The current generation of philosophers of space and time (led by Howard Stein and John Earman) have followed Einstein's lead here. But, interestingly, not only do these ideas boggle the mind of the average man in the street (like you and me), they boggle most contemporary scientific minds as well.[18] That is, I believe the majority opinion among working, knowledgeable scientists is that general relativity provides a magnificent organizing tool for treating certain gravitational problems in astrophysics and cosmology. But few, I believe, give credence to the kind of realist existence and nonexistence claims that I have been mentioning. For relativistic physics, then, it appears that a nonrealist attitude was important in its development, that the founder nevertheless espoused a realist attitude to the finished product, but that most who actually use it think of the theory as a powerful instrument, rather than as expressing a "big truth."

With quantum theory, this sequence gets a twist. Heisenberg's seminal paper of 1925 is prefaced by the following abstract, announcing, in effect, his philosophical stance: "In this paper an attempt will be made to obtain bases for a quantum-theoretical mechanics based exclusively on relations between quantities observable in principle."[19] In the body of the paper, Heisenberg not only rejects any reference to unobservables; he also moves away from the very idea that one should try to form any picture of a reality underlying his mechanics. To be sure, E. Schrödinger, the second father of quantum theory, seems originally to have had a vague picture of an underlying wavelike reality for his own equation. But he was quick to see the difficulties here and, just as quickly, although reluctantly, abandoned the attempt to interpolate any reference to reality.[20] These instrumentalist moves, away from a realist construal of the emerging quantum theory, were given particular force by Bohr's so-called "philosophy of complementarity"; and this nonrealist position was consolidated at the time of the famous Solvay conference, in October of 1927, and is firmly in place today. Such quantum nonrealism is part of what every graduate physicist learns and practices. It is the conceptual backdrop to all the brilliant successes in atomic, nuclear, and particle physics over the past fifty years. Physicists have learned to think about their theory in a highly nonrealist way, and doing just that has brought about the most marvelous predictive success in the history of science.

The war between Einstein, the realist, and Bohr, the nonrealist, over the interpretation of quantum theory was not, I believe, just a sideshow in physics, nor an idle intellectual exercise. It was an important endeavor undertaken by Bohr on behalf of the enterprise of physics as a progressive science. For Bohr believed (and this fear was shared by Heisenberg, A. Sommerfield, W. Pauli, and M. Born—and all the major players) that Einstein's realism, if taken seriously, would block the consolidation and articulation of the new physics and, thereby, stop the progress of science. They were afraid, in particular, that Einstein's realism would lead the next generation of the brightest and best students into scientific dead ends. Alfred Landé, for example, as a graduate student, was interested in spending some time in Berlin to sound out Einstein's ideas. His supervisor was Sommerfeld, and recalling this period, Landé writes

> The more pragmatic Sommerfeld . . . warned his students, one of them this writer, not to spend too much time on the hopeless task of "explaining" the quantum but rather to accept it as fundamental and help work out its consequences.[21]

The task of "explaining" the quantum, of course, is the realist program for identifying a reality underlying the formulas of the theory and thereby explaining the predicative success of the formulas as approximately true descriptions of this reality. It is this program that I have criticized in the first part of this paper, and this same program that the builders of quantum theory saw as a scientific dead end. Einstein knew perfectly well that the issue was joined right here. In the summer of 1935, he wrote to Schrödinger,

> The real problem is that physics is a kind of metaphysics; physics describes 'reality'. But we do not know what 'reality' is. We know it only through physical description. . . . But the Talmudic philosopher sniffs at 'reality', as at a frightening creature of the naive mind.[22]

By avoiding the bogey of an underlying reality, the "Talmudic" originators of quantum theory seem to have set subsequent generations on precisely the right path. Those inspired by realist ambitions have produced no predictively successful physics. Neither Einstein's conception of a unified field nor the ideas of the de Broglie group about pilot waves, nor the Bohm-inspired interest in hidden variables has made for scientific progress. To be sure, several philosophers of physics, including another Hilary Putnam, and myself, have fought a battle over the last decade to show that the quantum theory is at least consistent with some kind of underlying reality. I believe that Hilary has abandoned the cause, perhaps in part on account of the recent Bell-inequality problem over correlation experiments, a problem that van Fraassen calls "the charybdis of realism."[23] My own recent work in the area suggests that we may still be able to keep realism afloat in this whirlpool.[24] But the possibility (as I still see it) for a realist account of the quantum domain should not lead us away from appreciating the historical facts of the matter.

One can hardly doubt the importance of a nonrealist attitude for the development and practically infinite success of the quantum theory. Historical counterfactuals are always tricky, but the sterility of actual realist programs in this area at least suggests that Bohr and company were right in believing that the road to scientific progress here would have been blocked by realism. The founders of quantum theory never turned on the nonrealist attitude that served them so well. Perhaps that is because the central underlying theoretical device of quantum theory, the densities of a complex-valued and infinite-dimensional wave function, are even harder to take seriously than is the four-dimensional manifold of relativity. But now, there comes a most curious twist. For just as the practitioners of relativity, I have suggested, ignore the *realist* interpretation in favor of a more pragmatic attitude toward the space-time structure, the quantum physicists would appear to make a similar reversal and to forget their nonrealist history and allegiance when it comes time to talk about new discoveries.

Thus, anyone in the business will tell you about the exciting period, in the fall of 1974, when the particle group at Brookhaven, led by Samuel Ting, discovered the J particle, just as a Stanford team at the Stanford Linear Accelerator Center (SLAC), under Burton Richter, independently found a new particle they called "ψ". These turned out to be one and the same, the so-called ψ/J particle (Mass 3,098 MeV, Spin 1, Resonance 67 KeV, Strangeness 0). To explain this new entity, the theoreticians were led to introduce a new kind of quark, the so-called charmed quark. The ψ/J particle is then thought to be made up out of a charmed quark and an anticharmed quark, with their respective spins aligned. But if this is correct, then there ought to be other such pairs anti-aligned, or with variable spin alignments, and these ought to make up quite new observable particles. Such predictions from the charmed-quark model have turned out to be confirmed in various experiments.

In this example, I have been intentionally a bit more descriptive in order to convey the realist feel to the way scientists speak in this area. For I want to ask whether this is a return to realism or whether, instead, it can somehow be reconciled with a fundamentally nonrealist attitude.[25] I believe that the nonrealist option is correct, but I will not defend that answer here, however, because its defense involves the articulation of a compelling and viable form of nonrealism; and that is the task of the third (and final) section of this paper.

NONREALISM

Even if the realist happens to be a talented philosopher, I do not believe that, in his heart, he relies for his realism on the rather sophisticated form of abductive argument that I have examined and rejected in the first section of this paper, and which the history of twentieth-century physics shows to be fallacious. Rather, if his heart is like mine (and I *do* believe in a common nature), then I suggest that a more simple and homely sort of argument is what grips him. It is this, and I will put it in the first person. I certainly trust the evidence of my senses, on the whole, with regard to the existence and features of everyday objects. And I have similar confidence in the system of "check, double-check, triple-check" of scientific investigation, as well as the other safeguards built into the institutions of science. So, if the scientists tell me that there really are molecules, and atoms, and ψ/J particles and, who knows, maybe even quarks, then so be it. I trust them and, thus, must accept that there really are such things, with their attendant properties and relations. Moreover, if the instrumentalist (or some other member of the species "non-realistica") comes along to say that these entities, and their attendants, are just fictions (or the like), then I see no more reason to believe him than to believe that *he is* a fiction, made up (somehow) to do a job on me; which I do not believe. It seems, then, that I had better be a realist. One can summarize this homely and compelling line as follows: it is possible to accept the evidence of one's senses and to accept, *in the same way*, the confirmed results of science only for a realist; hence, I should be one (and so should you!).

What is it to accept the evidence of one's senses and, *in the same way*, to accept confirmed scientific theories? It is to take them into one's life as true, with all that implies concerning adjusting one's behavior, practical and theoretical, to accommodate these truths. Now, of course, there are truths, and truths. Some are more central to us and our lives, some less so. I might be mistaken about anything, but were I mistaken about where I am right now, that might affect me more than would my perhaps mistaken belief in charmed quarks. Thus, it is compatible with the homely line of argument that some of the scientific beliefs that I hold are less central than some, for example, perceptual beliefs. Of course, were I deeply in the charmed-quark business, giving up that belief might be more difficult than giving up some at the perceptual level. (Thus we get the phenomenon of "seeing what you believe," as is well known to all thoughtful people.) When the homely line asks us, then, to accept the scientific results "in the same way" in which we accept the evidence of our senses, I take it that we are to accept them both as true. I take it that we are being asked not to distinguish between kinds of truth or modes of existence or the like, but only among truths themselves, in terms of centrality, degrees of belief, or such.

Let us suppose this understood. Now, do you think that Bohr, the archenemy of realism, could toe the homely line? Could Bohr, fighting for the sake of science (against Einstein's realism) have felt compelled either to give up the results of science, or else to assign to its "truths" some category different from the truths of everyday life? It seems unlikely. And thus, unless we uncharitably think Bohr inconsistent on this basic issue, we might well come to question whether there is any necessary connection moving us from accepting the results of science as true to being a realist.[26]

Let me use the term 'antirealist' to refer to any of the many different specific enemies of realism: the idealist, the instrumentalist, the phenomenalist, the empiricist (constructive or not), the conventionalist, the constructivist, the pragmatist, and so forth. Then, it seems to me that both the realist and the antirealist must toe what I have been calling "the homely line." That is, they must both accept the certified results of science as on par with more homely and familiarly supported claims. That is not to say that one party (or the other) cannot distinguish more from less well-confirmed claims at home or in science; nor that one cannot single out some particular mode of inference (such as inference to the best explanation) and worry over its reliability, both at home and away. It is just that one must maintain parity. Let us say, then, that both realist and antirealist accept the results of scientific investigations as 'true', on par with more homely truths. (I realize that some antirealists would rather use a different word, but no matter.) And call this acceptance of scientific truths the "core position."[27] What distinguishes realists from antirealists, then, is what they add onto this core position.

The antirealist may add onto the core position a particular analysis of the concept of truth, as in the pragmatic and instrumentalist and conventionalist conceptions of truth. Or the antirealist may add on a special analysis of concepts, as in idealism, constructivism, phenomenalism, and in some varieties of empiricism. These addenda will then issue in a special meaning, say, for existence statements. Or the antirealist may add on certain methodological strictures, pointing a wary finger at some particular inferential tool, or constructing his own account for some particular aspects of science (e.g., explanations or laws). Typically, the antirealist will make several such additions to the core.

What then of the realist, what does he add to his core acceptance of the results of science as really true? My colleague, Charles Chastain, suggested what I think is the most graphic way of stating the answer—namely, that what the realist adds on is a desk-thumping, foot-stamping shout of "Really!" So, when the realist and antirealist agree, say, that there really are electrons and that they really carry a unit negative charge and really do have a small mass (of about 9.1×10^{-28} grams), what the realist wants to add is the emphasis that all this is really so. "There really are electrons, really!" This typical realist emphasis serves both a negative and a positive function. Negatively, it is meant to deny the additions that the antirealist would make to that core acceptance which both parties share. The realist wants to deny, for example, the phenomenalistic reduction of concepts or the pragmatic conception of truth. The realist thinks that these addenda take away from the substantiality of the accepted claims to truth or existence. "No," says he, "they *really* exist, and not in just your diminished antirealist sense." Positively, the realist wants to explain the robust sense in which *he* takes these claims to truth or existence, namely, as claims about reality—what is really, really the case. The full-blown version of this involves the conception of truth as correspondence with the world, and the surrogate use of approximate truth as near-correspondence. We have already seen how these ideas of correspondence and approximate truth are supposed to explain what *makes* the truth *true* whereas, in fact, they function as mere trappings, that is, as superficial decorations that may well attract our attention but do not compel rational belief. Like the extra "really," they are an arresting foot-thump and, logically speaking, of no more force.

It seems to me that when we contrast the realist and the antirealist in terms of what they each want to add to the core position, a third alternative emerges—and an attractive one at that. It is the core position itself, *and all by itself.* If I am correct in thinking that, at heart, the grip of realism only extends to the homely connection of everyday truths with scientific truths, and that good sense dictates our acceptance of the one on the same basis as our acceptance of the other, then the homely line makes the core position, all by itself, a compelling one, one that we ought to take to heart. Let us try to do so, and to see whether it constitutes a philosophy, and an attitude toward science, that we can live by.

The core position is neither realist nor antirealist; it mediates between the two. It would be nice to have a name for this position, but it would be a shame to appropriate another "ism" on its behalf, for then it would appear to be just one of the many contenders for ontological allegiance. I think it is not just one of that crowd but rather, as the homely line behind it suggests, it is for commonsense epistemology—the natural ontological attitude. Thus, let me introduce the acronym *NOA* (pronounced as in "Noah"), for *natural ontological attitude*, and, henceforth, refer to the core position under that designation.

To begin showing how NOA makes for an adequate philosophical stance toward science, let us see what it has to say about ontology. When NOA counsels us to accept the results of science as true, I take it that we are to treat truth in the usual referential way, so that a sentence (or statement) is true just in case the entities referred to stand in the referred-to relations. Thus, NOA sanctions ordinary referential semantics and commits us, via truth, to the existence of the individuals, properties, relations, processes, and so forth referred to by the scientific statements that we accept as true. Our belief in their existence will be just as strong (or weak) as our belief in the truth of the bit of science involved, and degrees of belief here, presumably, will be tutored by ordinary relations of confirmation and evidential support, subject to the usual scientific canons. In taking this referential stance, NOA is not committed to the progressivism that seems inherent in realism. For the realist, as an article of faith, sees scientific success, over the long run, as bringing us closer to the truth. His whole explanatory enterprise, using approximate truth, forces his hand in this way. But, a "noaer" (pronounced as "knower") is not so committed. As a scientist, say, within the context of the tradition in which he works, the noaer, of course, will believe in the existence of those entities to which his theories refer. But should the tradition change, say in the manner of the conceptual revolutions that Kuhn dubs "paradigm shifts," then nothing in NOA dictates that the change be assimilated as being progressive, that is, as a change where we learn more accurately about *the same things*. NOA is perfectly consistent with the Kuhnian alternative, which construes such changes as wholesale changes of reference. Unlike the realist, adherents to NOA are free to examine the facts in cases of paradigm shift, and to see whether or not a convincing case for stability of reference across paradigms can be made without superimposing on these facts a realist-progressivist superstructure. I have argued elsewhere that if one makes oneself free, as NOA enables one to do, then the facts of the matter will not usually settle the case;[28] and that this is a good reason for thinking that cases of so-called "incommensurability" are, in fact, genuine cases where the question of stability of reference is indeterminate. NOA, I think, is the right philosophical position for such conclusions. It sanctions reference and existence claims, but it does not force the history of science into prefit molds.

So far I have managed to avoid what, for the realist, is the essential point, for what of the "external world"? How can I talk of reference and of existence claims unless I am talking about referring to things right out there in the world? And here, of course, the realist, again, wants to stamp his feet.[29] I think the problem that makes the realist want to stamp his feet, shouting "Really!" (and invoking the external world) has to do with the stance the realist tries to take vis-à-vis the game of science. The realist, as it were, tries to stand outside the arena watching the ongoing game and then tries to judge (from this external point of view) what the point is. It is, he says, *about* some area external to the game. The realist, I think, is fooling himself. For he cannot (really!) stand outside the arena, nor can he survey some area off the playing field and mark it out as what the game is about.

Let me try to address these two points. How are we to arrive at the judgment that, in addition to, say, having a rather small mass, electrons are objects "out there in the external world"? Certainly, we can stand off from the electron game and survey its claims, methods, predictive success, and so forth. But what stance could we take that would enable us to judge what the theory of electrons is *about*, other than agreeing that it is about electrons? It is not like matching a blueprint to a house being built, or a map route to a country road. For we are *in* the world, both physically and conceptually.[30] That is, *we* are among the objects of science, and the concepts and procedures that we use to make judgments of subject matter and correct application are themselves part of that same scientific world. Epistemologically, the situation is very much like the situation with regard to the justification of induction. For the problem of the external world (so-called) is how to satisfy the realist's demand that we justify the existence claims sanctioned by science (and, therefore, by NOA) as claims to the existence of entities "out there." In the case of induction, it is clear that only an inductive justification will do, and it is equally clear that no inductive justification will do at all. So too with the external world, for only ordinary scientific inferences to existence will do, and yet none of them satisfies the demand for showing that the existent is really "out there." I think we ought to follow Hume's prescription on induction, with regard to the external world. There is no possibility for justifying the kind of externality that realism requires, yet it may well be that, in fact, we cannot help yearning for just such a comforting grip on reality. I shall return to this theme at the close of the paper.

If I am right, then the realist is chasing a phantom, and we cannot actually do more, with regard to existence claims, than follow scientific practice, just as NOA suggests. What then of the other challenges raised by realism? Can we find in NOA the resources for understanding scientific practice? In particular (since it was the topic of the first part of this paper), does NOA help us to understand the scientific method, such as the problems of the small handful or of conjunctions? The sticking point with the small handful was to account for why the few and narrow alternatives that we can come up with, result in successful novel predictions, and the like. The background was to keep in mind that most such narrow alternatives are not successful. I think that NOA has only this to say. If you believe that guessing based on some truths is more likely to succeed than guessing pure and simple, then if our earlier theories were in large part true and if our refinements of them conserve the true parts, then guessing on this basis has some relative likelihood of success. I think this is a weak account, but then I think the phenomenon here does not allow for anything much stronger since, for the most part, such guesswork fails. In the same way, NOA can help with the problem of conjunctions (and, more generally, with problems of logical combinations). For if two consistent theories in fact have overlapping domains (a fact, as I have just suggested, that is not so often decidable), and if the theories also have true things to say about members in the overlap, then conjoining the theories just adds to the truths of each and, thus, *may*, in conjunction, yield new truths. Where one finds other successful methodological rules, I think we will find NOA's grip on the truth sufficient to account for the utility of the rules.

Unlike the realist, however, I would not tout NOA's success at making science fairly intelligible as an argument in its favor, vis-à-vis realism or various antirealisms. For NOA's accounts are available to these fellows, too, provided what they add to NOA does not negate its appeal to the truth, as does a verificationist account of truth or the realists' longing for approximate truth. Moreover, as I made plain enough in the first section of this paper, I am sensitive to the possibility that explanatory efficacy can be achieved without the explanatory hypothesis being true. NOA may well make science seem fairly intelligible and even rational, but NOA could be quite the wrong view of science for all that. If we posit as a constraint on philosophizing about science that the scientific enterprise should come out in our philosophy as not too unintelligible or irrational, then, perhaps, we can say that NOA passes a minimal standard for a philosophy of science.

Indeed, perhaps the greatest virtue of NOA is to call attention to just how minimal an adequate philosophy of science can be. (In this respect, NOA might be compared to the minimalist movement in art.) For example, NOA helps us to see that realism differs from various antirealisms in this way: realism adds an outer direction to NOA, that is, the external world and the correspondence relation of approximate truth; antirealisms (typically) add an inner direction, that is, human-oriented reductions of truth, or concepts, or explanations (as in my opening citation from Hume). NOA suggests that the legitimate features of these additions are already contained in the presumed equal status of everyday truths with scientific ones, and in our accepting them both as *truths*. No other additions are legitimate, and none are required.

It will be apparent by now that a distinctive feature of NOA, one that separates it from similar views currently in the air, is NOA's stubborn refusal to amplify the concept of truth, by providing a theory or analysis (or even a metaphorical picture). Rather, NOA recognizes in "truth" a concept already in use and agrees to abide by the standard rules of usage. These rules involve a Davidsonian-Tarskian, referential semantics, and they support a thoroughly classical logic of inference. Thus NOA respects the customary "grammar" of 'truth' (and its cognates). Likewise, NOA respects the customary epistemology, which grounds judgments of truth in perceptual judgments and various confirmation relations. As with the use of other concepts, disagreements are bound to arise over what is true (for instance, as to whether inference to the best explanation is always truth-conferring). NOA pretends to no resources for settling these disputes, for NOA takes to heart the great lesson of twentieth-century analytic and Continental philosophy, namely, that there *are* no general methodological or philosophical resources for deciding such things. The mistake common to realism and all the antirealisms alike is their commitment to the existence of such nonexistent resources. If pressed to answer the question of what, then, does it *mean* to say that something is true (or to what does the truth of so-and-so commit one), NOA will reply by pointing out the logical relations engendered by the specific claim and by focusing, then, on the concrete historical circumstances that ground that particular judgment of truth. For, after all, there *is* nothing more to say.[31]

Because of its parsimony, I think the minimalist stance represented by NOA marks a revolutionary approach to understanding science. It is, I would suggest, as profound in its own way as was the revolution in our conception of morality, when we came to see that founding morality on God and His Order was *also* neither legitimate nor necessary. Just as the typical theological moralist of the eighteenth century would feel bereft to read, say, the pages of *Ethics*, so I think the realist must feel similarly when NOA removes that "correspondence to the external world" for which he so longs. I too have regret for that lost paradise, and too often slip into the realist fantasy. I use my understanding of twentieth-century physics to help me firm up my convictions about NOA, and I recall some words of Mach, which I offer as a comfort and as a closing. With reference to realism, Mach writes

> It has arisen in the process of immeasurable time without the intentional assistance of man. It is a product of nature, and preserved by nature. Everything that philosophy has accomplished . . . is, as compared with it, but an insignificant and ephemeral product of art. The fact is, every thinker, every philosopher, the moment he is forced to abandon his one-sided intellectual occupation . . . , immediately returns [to realism].
>
> Nor is it the purpose of these "introductory remarks" to discredit the standpoint [of realism]. The task which we have set ourselves is simply to show why and for what purpose we hold that standpoint during most of our lives, and why and for what purpose we are . . . obliged to abandon it.

These lines are taken from Mach's *The Analysis of Sensations* (Sec. 14). I recommend that book as effective realism-therapy, a therapy that works best (as Mach suggests) when accompanied by historicophysical investigations (real versions of the breakneck history of my second section, "Realism and Progress"). For a better philosophy, however, I recommend NOA.

NOTES

My thanks to Charles Chastain, Gerald Dworkin, and Paul Teller for useful preliminary conversations about realism and its rivals, but especially to Charles— for only he, then, (mostly) agreed with me, and surely that deserves special mention. This paper was written by me, but cothought by Micky Forbes. I don't know any longer whose ideas are whose. That means that the responsibility for errors and confusions is at least half Micky's (and she is two-thirds responsible for "NOA"). Finally, I am grateful to the many people who offered comments and criticisms at the conference, and subsequently. I am also grateful to the National Science Foundation for a grant in support of this research.

1. In the final section, I call this postrealism "NOA." Among recent views that relate to NOA, I would include Hilary Putnam's "internal realism," Richard Rorty's "epistemological behaviorism," the "semantic realism" espoused by Paul Horwich, parts of the "Mother Nature" story told by William Lycan, and the defense of common sense worked out by Joseph Pitt (as a way of reconciling W. Sellars's manifest and scientific images). For references, see Hilary Putnam,

Meaning and the Moral Sciences (London: Routledge and Kegan Paul, 1978); Richard Rorty, *Philosophy and the Mirror of Nature* (Princeton: Princeton University Press, 1979); Paul Horwich, "Three Forms of Realism," *Synthese* 51 (1982): 181-201; William G. Lycan, "Epistemic Value" (preprint, 1982); and Joseph C. Pitt, *Pictures, Images and Conceptual Change* (Dordrecht: D. Reidel, 1981). The reader will note that some of the above consider their views a species of realism, whereas others consider their views antirealist. As explained below, NOA marks the divide; hence its "postrealism."

2. Larry Laudan, "A Confutation of Convergent Realism," this volume.

3. Richard N. Boyd, "Scientific Realism and Naturalistic Epistemology," in *PSA* (1980), vol. 2, ed. P. D. Asquith and R. N. Giere (E. Lansing: Philosophy of Science Association, 1981), 613-662. See also, Boyd's article in this book, and further references there.

4. Hilary Putnam, "The Meaning of 'Meaning'," in *Language, Mind and Knowledge*, ed. K. Gunderson (Minneapolis: University of Minnesota Press, 1975), 131-193. See also his article in this volume.

5. Bas C. van Fraasen, *The Scientific Image* (Oxford: The Clarendon Press, 1980). See especially pp. 97-101 for a discussion of the truth of explanatory theories. To see that the recent discussion of realism is joined right here, one should contrast van Fraassen with W. H. Newton-Smith, *The Rationality of Science* (London: Routledge and Kegan Paul, 1981), esp. chap. 8.

6. Nancy Cartwright's *How The Laws of Physics Lie* (Oxford: Oxford University Press, 1983) includes some marvelous essays on these issues.

7. Some realists may look for genuine support, and not just solace, in such a coherentist line. They may see in their realism a basis for general epistemology, philosophy of language, and so forth (as does Boyd, "Scientific Realism and Naturalistic Epistemology"). If they find in all this a coherent and comprehensive world view, then they might want to argue for their philosophy as Wilhelm Wien argued (in 1909) for special relativity, "What speaks for it most of all is the inner consistency which makes it possible to lay a foundation having no self-contradictions, one that applies to the totality of physical appearances." Quoted by Gerald Holton, "Einstein's Scientific Program: Formative Years" in *Some Strangeness in the Proportion*, ed. H. Woolf (Reading: Addison-Wesley, 1980), 58. Insofar as the realist moves away from the abductive defense of realism to seek support, instead, from the merits of a comprehensive philosophical system with a realist core, he marks as a failure the bulk of recent defenses of realism. Even so, he will not avoid the critique pursued in the text. For although my argument above has been directed, in particular, against the abductive strategy, it is itself based on a more general maxim, namely, that the form of argument used to support realism must be more stringent than the form of argument embedded in the very scientific practice that realism itself is supposed to ground—on pain of begging the question. Just as the abductive strategy fails because it violates this maxim, so too would the coherentist strategy, should the realist turn from one to the other. For, as we see from the words of Wien, the same coherentist line that the realist would appropriate for his own support, is part of ordinary scientific practice in framing judgments about competing theories. It is, therefore, not a line of defense avail-

able to the realist. Moreover, just as the truth-bearing status of abduction is an issue dividing realists from various nonrealists, so too is the status of coherence-based inference. Turning from abduction to coherence, therefore, still leaves the realist begging the question. Thus, when we bring out into the open the character of arguments *for* realism, we see quite plainly that they do not work.

In support of realism there seem to be only those "reasons of the heart" which, as Pascal says, reason does not know. Indeed, I have long felt that belief in realism involves a profound leap of faith, not at all dissimilar from the faith that animates deep religious convictions. I would welcome engagement with realists on this understanding, just as I enjoy conversation on a similar basis with my religious friends. The dialogue will proceed more fruitfully, I think, when the realists finally stop pretending to a rational support for their faith, which they do not have. Then we can all enjoy their intricate and sometimes beautiful philosophical constructions (of, e.g., knowledge, or reference, etc), even though, as nonbelievers, they may seem to us only wonderful castles in the air.

8. I hope all readers of this essay will take this idea to heart. For in formulating the question as how to explain why the methods of science lead to instrumental success, the realist has seriously misstated the explanandum. Overwhelmingly, the results of the conscientious pursuit of scientific inquiry are failures: failed theories, failed hypotheses, failed conjectures, inaccurate measurements, incorrect estimations of parameters, fallacious causal inferences, and so forth. If explanations are appropriate here, then what requires explaining is why the very same methods produce an overwhelming background of failures and, occasionally, also a pattern of successes. The realist literature has not yet begun to address this question, much less to offer even a hint of how to answer it.

9. Of course, the realist can appropriate the devices and answers of the instrumentalist, but that would be cheating, and it would, anyway, not provide the desired support of realism per se.

10. Paul Teller has made this suggestion to me in conversation.

11. Ilkka Niiniluoto's "What Shall We Do with Verisimilitude?" *Philosophy of Science* 49 (1982): 181-197, contains interesting formal constructions for "degree of truthlikeness," and related versimilia. As conjectured above, they rely on an unspecified correspondence relation to the truth and on measures of the "distance" from the truth. Moreover, they fail to sanction that projection from some approximate truths to other, novel truths, which lies at the core of realist rationalizations.

12. See Gerald Holton, "Mach, Einstein, and the Search for Reality," in his *Thematic Origins of Scientific Thought* (Cambridge: Harvard University Press, 1973), 219-259. I have tried to work out the precise role of this positivist methodology in my "The Young Einstein and the Old Einstein," in *Essays in Memory of Imré Lakatos*, ed. R. S. Cohen et al. (Dordrecht: D. Reidel, 1976), 145-159.

13. A. Einstein et al., *The Principle of Relativity*, trans. W. Perrett and G. B. Jeffrey (New York: Dover, 1952), 117.

14. John Earman et al., "Lost in the Tensors," *Studies in History and Philosophy of Science* 9 (1978): 251-278. The tortuous path detailed by Earman is sketched by B. Hoffmann, *Albert Einstein, Creator and Rebel* (New York: New

American Library, 1972), 116-128. A nontechnical and illuminating account is given by John Stachel, "The Genesis of General Relativity," in *Einstein Symposium Berlin*, ed. H. Nelkowski et al. (Berlin: Springer-Verlag, 1980).

15. I have in mind the role played by the analysis of simultaneity in Einstein's path to special relativity. Despite the important study by Arthur Miller, *Albert Einstein's Special Theory of Relativity* (Reading: Addison-Wesley, 1981), and an imaginative pioneering work by John Earman (and collaborators), the details of which I have been forbidden to disclose, I think the role of positivist analysis in the 1905 paper has yet to be properly understood. My ideas here were sparked by Earman's playful reconstructions. So I cannot expose my ideas until John is ready to expose his.

16. Peter Barker, "Einstein's Later Philosophy of Science," in *After Einstein*, ed. P. Barker and C. G. Shugart (Memphis: Memphis State University Press, 1981), 133-146, is a nice telling of this story.

17. Roger Jones in "Realism About What?" (in draft) explains very nicely some of the difficulties here.

18. I think the ordinary, deflationist attitude of working scientists is much like that of Steven Weinberg, *Gravitation and Cosmology: Principles and Applications of the General Theory of Relativity* (New York: Wiley, 1972).

19. See B. L. van der Waerden, *Sources of Quantum Mechanics* (New York: Dover, 1967), 261.

20. See Linda Wessels, "Schrödinger's Route to Wave Mechanics," *Studies in History and Philosophy of Science* 10 (1979): 311-340.

21. A. Landé, "Albert Einstein and the Quantum Riddle," *American Journal of Physics* 42 (1974): 460.

22. Letter to Schrödinger, June 19, 1935. See my "Einstein's Critique of Quantum Theory: The Roots and Significance of EPR," in *After Einstein* (see n. 16), 147-158, for a fuller discussion of the contents of this letter.

23. Bas van Fraassen, "The Charybdis of Realism: Epistemological Implications of Bell's Inequality," *Synthese* 52 (1982): 25-38.

24. See my "Antinomies of Entanglement: The Puzzling Case of The Tangled Statistics," *Journal of Philosophy* 79, 12 (1982), for part of the discussion and for reference to other recent work.

25. The nonrealism that I attribute to students and practitioners of the quantum theory requires more discussion and distinguishing of cases and kinds than I have room for here. It is certainly not the all-or-nothing affair I make it appear in the text. I hope to carry out some of the required discussion in a talk for the American Philosophical Association meetings, Pacific Division, March 1983, entitled "Is Scientific Realism Compatible with Quantum Physics?" My thanks to Paul Teller and James Cushing, each of whom saw the need for more discussion here.

26. I should be a little more careful about the historical Bohr than I am in the text. For Bohr himself would seem to have wanted to truncate the homely line somewhere between the domain of chairs and tables and atoms, whose existence he plainly accepted, and that of electrons, where he seems to have thought the question of existence (and of realism, more generally) was no longer well defined.

An illuminating and provocative discussion of Bohr's attitude toward realism is given by Paul Teller, "The Projection Postulate and Bohr's Interpretation of Quantum Mechanics," pp. 201-223 n. 3. Thanks, again, to Paul for helping to keep me honest.

22. In this context, for example, van Fraassen's "constructive empiricism" would prefer the concept of empirical adequacy, reserving "truth" for an (unspecified) literal interpretation and believing in that truth only among observables. It is clear, nevertheless, that constructive empiricism follows the homely line and accepts the core position. Indeed, this seems to be its primary motivating rationale. If we reread constructive empiricism in our terminology, then, we would say that it accepts the core position but adds to it a construal of truth as empirical adequacy. Thus, it is antirealist, just as suggested in the next paragraph below. I might mention here that in this classification Putnam's internal realism also comes out as antirealist. For Putnam also accepts the core position, but he would add to it a Peircean construal of truth as ideal rational acceptance. This is a mistake, which I expect that Putnam will realize and correct in future writings. He is criticized for it, soundly I think, by Paul Horwich ("Three Forms of Realism") whose own "semantic realism" turns out, in my classification, to be neither realist nor antirealist. Indeed, Horwich's views are quite similar to what is called "NOA" below, and could easily be read as sketching the philosophy of language most compatible with NOA. Finally, the "epistemological behaviorism" espoused by Rorty is a form of antirealism that seems to me very similar to Putnam's position, but achieving the core parity between science and common sense by means of an acceptance that is neither ideal nor especially rational, at least in the normative sense. (I beg the reader's indulgence over this summary treatment of complex and important positions. I have been responding to Nancy Cartwright's request to differentiate these recent views from NOA. So if the treatment above strikes you as insensitive, or boring, please blame Nancy.)

28. "How To Compare Theories: Reference and Change," *Nous* 9 (1975): 17-32.

29. In his remarks at the Greensboro conference, my commentator, John King, suggested a compelling reason to prefer NOA over realism; namely, because NOA is less percussive! My thanks to John for this nifty idea, as well as for other comments.

30. "There is, I think, no theory-independent way to reconstruct phrases like 'really true'; the notion of match between the ontology of a theory and its 'real' counterpart in nature now seems to me illusive in principle." T. S. Kuhn, "Postscript," in *The Structure of Scientific Revolutions*, 2d ed. (Chicago: University of Chicago Press, 1970), 206. The same passage is cited for rebuttal by W. H. Newton-Smith, in *The Rationality of Science*. But the "rebuttal" sketched there in chapter 8, sections 4 and 5, not only runs afoul of the objections stated here in my first section, it also fails to provide for the required theory-independence. For Newton-Smith's explication of verisimilitude (p. 204) makes explicit reference to some unspecified background theory. (He offers either current science or the Peircean limit as candidates.) But this is not to rebut Kuhn's challenge (and mine); it is to concede its force.

31. No doubt I am optimistic, for one can always think of more to say. In particular, one could try to fashion a general, descriptive framework for codifying and classifying such answers. Perhaps there would be something to be learned from such a descriptive, semantical framework. But what I am afraid of is that this enterprise, once launched, would lead to a proliferation of frameworks not so carefully descriptive. These would take on a life of their own, each pretending to ways (better than its rivals) to settle disputes over truth claims, or their import. What we need, however, is less bad philosophy, not more. So here, I believe, silence is indeed golden.

5

The Path from Data to Theory

Ronald Laymon

Philosophers tend to view the relationship between a particular scientific theory and its data as representable by means of a single logical structure. The relationship is not ordinarily viewed as requiring a hierarchy of structures of different logical types. Two excellent books, recently published, illustrate this tendency. Clark Glymour, in *Theory and Evidence*,[1] continues the positivist tradition of viewing theories as axiomatic systems which connect with their evidence by means of ordinary quantification and truth-functional connectives. Theory testing can be understood in terms of the syntactic properties of a single logical system. In *The Scientific Image*,[2] van Fraassen adopts a version of what is sometimes called the "semantic view": a theory is a set of models of some purely formal system. Evidence constitutes (if confirmatory) a submodel embedded in a model of the equivalence class that is the theory. As on the axiomatic view, the relationship between evidence and theory is analyzable in terms of the formal properties of a single logical system. Of course, canonical representation of these sorts is usually admitted to be an idealization or first approximation of actual practice.

No one would deny that in actual practice the road from data to theory is long and arduous. There are complications owing to the need for auxiliary theories of instrumentation, for justifications of particular measures of goodness of fit, and for reasons establishing that the relevant *ceteris paribus* clauses have been isolated and satisfied. Sometimes the theory to be tested must be used as the theory of the testing instruments. The general question to raise here is: does the idealization of assuming a canonical representation of single logical type (for each theory) seriously distort our understanding of the nature of science? In this paper, I shall consider the specific issue of the effect a more realistic analysis of data and theory may have on argumentation for and against scientific realism.[3]

Arguments against realism typically begin with some canonical or near-canonical representation of scientific theory. This representation is then sometimes transformed in a way that avoids nonobservable theoretical entities as, for example, by transforming axioms into their Craig or Ramsey equivalents. Another move is simply to deny a semantics for nonobservables. It is then argued that we accept the transformed or restricted canonical form on the basis of Occam's razor: *ceteris paribus* choose canonical representations committed to as few entities as possible. Advocates of realism respond by trying to show that the *ceteris paribus* clause is not satisfied when comparing nontransformed or semantically complete representations with their antirealist instrumentalist brethren. If, though, our initial canonical representation is modified to reflect a more complex connection between data and theory, then it is possible that Occam's razor will cut differently and that the possibilities for *ceteris paribus* violations will be greatly changed.

The starlight deflection experimental tests of the general theory of relativity will be sufficient to illustrate the type of complexity I have in mind. To apply the field equations, one must specify the mass-energy distribution of the particular system in question. However, even accepting Newtonian measures of the sun and earth, the field equations cannot be brought to bear since their solution, given such a realistic description, is unknown and probably computationally impossible. So an *idealized description* is employed: the Schwarzschild "solution" assumes a perfectly symmetrical nonrotating sun and no other masses. Together the field equations and the Schwarzschild idealization yield a solution for the metric.[4]

Before proceeding farther in the description of this case, let me isolate a philosophically interesting feature. Strictly speaking, the Schwarzschild description of sun and gravitational field as symmetric and static is false. If this description is viewed as an initial condition, the predictive inference is therefore unsound. Hence, no conclusion about the truth values of the premises is deducible. The usual response to such an observation is to object that the Schwarzschild description is approximately true or an adequate idealization. This cannot be denied. However, this piece of ordinary scientific wisdom is no philosophical analysis of these concepts, nor does it specify what justifies the claim of adequacy or approximate truth. A useful first step toward a philosophical theory of adequate idealizations is to view them as intended antecedents for some species of *counterfactual*: if the sun were static and symmetrical, then it would deflect starlight by such and such amount. An interest relative possible-world interpretation seems natural here where truth is controlled by (among other things) the field equations. Adequacy of idealization corresponds, then, to some concept of sufficient "nearness" of the possible world to our world. Regardless of how the philosophical analysis goes, there must be some justification for regarding the Schwarzschild solution as appropriate for the deflection experiment.

The next step in the derivation of an observable prediction is to compute the constraints on orbital shapes imposed by the metric. The details of this computation need not concern us here. What is of interest is how the orbital equation, expressed in rather abstract theoretical terms, gets attached to our measuring instruments. What we read in standard textbooks is the following: if we assume that the metric "at the points of origin and detection" of deflected photons is Minkowskian, and if we ignore complications owing to quantum effects, then "there is no question about the meaning of" our relativistic computation of orbit: "it is the azimuthal angle in a system of coordinates within which light rays define lines that are essentially straight." Hence, we can "relate" our computation "to the shift of stellar images on photographic plates by the ordinary rules of geometric optics."[5] In effect, the relativistic description is *transformed* into a simple Euclidean calculation of the deflection of starlight compared with what it *would have been if* the sun were not present. Thinking in terms of our proposed possible-worlds analysis, what we have is some sort of mapping from a set of possible worlds controlled by the field equations to a set of worlds controlled now by Euclidean geometry.

There is an additional complication. The purpose of the eclipse is to enable us to determine stellar positions ordinarily invisible to us. Such starlight is bent by the sun. After eclipse, these stars are once again invisible and are not observable in the night sky until several months later. Hence, there are small changes in observational conditions that need to be corrected for. Differences due, for example, to parallax and the earth's motion are calculated out by means of Euclidean geometry, in order to test the intended comparison (see fig. 1).

From the point of view of the data collector, the procedure determined by the above analysis and reduction is to photograph stellar fields at the time of eclipse and again several months later when these fields are visible in the evening sky. The photographic plates are then to be superimposed and relative displacements noted. These displacements should provide the desired experimental test. And so they would if it were not for (among other things) scale distortion caused by small immeasurable

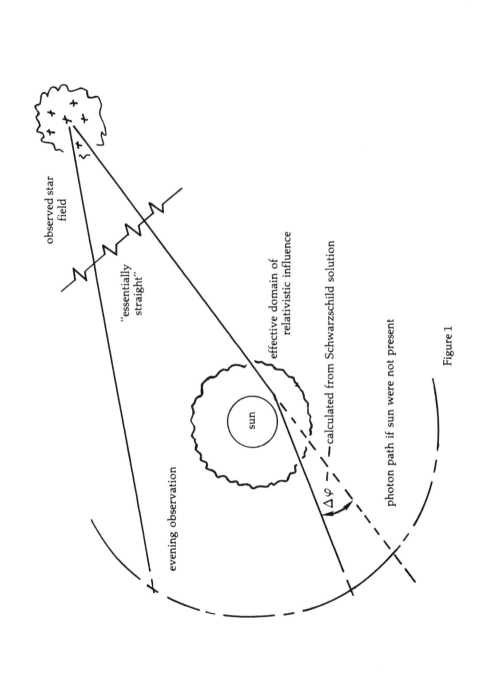

observed star
field

"essentially
straight"

observed star field

evening observation

sun

effective domain of
relativistic influence

$\Delta \varphi$

calculated from Schwarzschild solution

photon path if sun were not present

Figure 1

changes in the optical system as it waits the required several months. The basic means of overcoming this distortion problem is to test the prediction using a weighed least-squares measure of goodness of fit. Particular measures used are justified, in part, by appeal to a Euclidean theory of the measuring instruments. Since the theoretical prediction is of a hyperbolic dependence between displacement and distance from the sun, what is sought for is the hyperbola that fits the data best, assuming a normal distribution of errors and a relative weighing of error type importance. (An interesting feature of this case is that because of the absence of data at distances close to the sun, straight lines can be made to fit the data as well as can hyperbolas!) (See fig. 2.) Actually, we are not done yet, since there are implicit *ceteris paribus* clauses in the measures used of goodness of fit. For example, seriously distorted star images are assumed indicative of some systematic error in violation of the goodness of fit randomization assumptions.

While there are individual differences, I believe that the general level of complexity illustrated here, on the road from theory to data, is typical of the experimental testing of scientific theories. Let us assume that it is. Our somewhat impressionistic rendering yields this hierarchy of logical structures.[6]

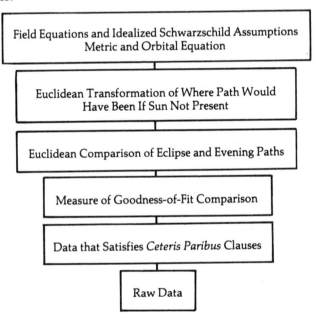

The problem of this paper is to determine what effect consideration of this hierarchical counterfactual structure leading from data to theory will

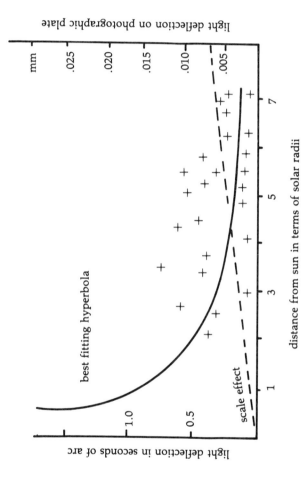

light deflection on photographic plate

mm

.025

.020

.015

.010

.005

best fitting hyperbola

scale effect

distance from sun in terms of solar radii

1   3   5   7

1.0

0.5

light deflection in seconds of arc

Scale distortion calculated for alteration of 0.1 mm in focal setting for 343 cm focal length

Figure 2

have on the issue of scientific realism. Prima facie, the case for instrumentalism seems greatly strengthened. After all, everything is treated counterfactually and knowingly so. Therefore, science does not aim to produce "true theories," but instead opts for calculational convenience and empirical adequacy. This *is* true. Such an admission, however, need not stand in the way of realism. Let me explain.

The antirealist will want to distinguish between two kinds of idealizations. First, there is an idealized treatment of what the antirealist antecedently accepts as an actual object, usually, something humanly observable. Treating the sun or the earth as perfectly spherical and nonrotating is an example. A theory of *reference* that ties the idealized *description* that will be "true" of some object in some possible world to that "same" object in this world seems a reasonable, and by now traditional, way to proceed. The antirealist will also want to distinguish a second kind of idealization, namely, idealized treatments of theoretical nonobservable objects. But here, the antirealist will turn instrumentalist and deny reference for the apparent object terms. Since ordinary ostension does not do duty here, nothing of value seems lost by giving up the advantages of a referential account. The problem of explaining the sense in which a theory of unobservables can be said to be idealized remains. If unobservables do not exist, how can descriptions of them be idealized? Consider J. Jeans's simple mean free path version of kinetic theory. While idealized, it is less idealized than simple collisional theories. For example, Jeans describes the case this way:

the pressure of a gas was calculated on the assumptions that the molecules were infinitesimal in size, and that they exerted no forces on one another except when they were actually in collision. . . . Neither of these assumptions is true for an actual gas in which the molecules are of finite size, and exert forces of cohesion on one another even when they are not in contact.[7]

The surface grammar here seems to demand a semantic account in terms of more realistic descriptions of existing objects. Before considering the antirealist options, allow me some additional stage setting. In particular, we will need some consequences for *confirmation* that can be drawn from the hierarchical counterfactual structure connecting data and theory. Obviously, there is a close connection between our concepts of confirmation and of the existence of theoretical objects.

Above, I asked the question: what justifies the idealizations used in science? The claim that it is the *success* of the resultant predictions will not do, since theories never exactly predict the phenomena. Now, it might be claimed that this failure of exact predictive fit can be accommodated within statistical theories of goodness of fit. While this is certainly part of the answer, it must be remembered that, typically, idealized theories of the measuring instruments are used to justify the employed measure of goodness of fit. In the starlight bending experiments, simple Euclidean analyses were given to justify data correction and the identification and extent of factors whose random variation would affect experimental outcome. The adequacy of such analyses gets justified on grounds such as: (1) "at infinity the metric becomes 'Minkowskian'," and (2) the gravitational field "on the earth's surface is more than 10^3 times weaker than that of the sun on the sun's surface."[8] Explicit arguments to demonstrate the cogency of such grounds are rarely, if ever, given. The traditional folklore of a scientific education says that such approximations are adequate. My suspicion is that, if pressed, the best that could be done, in the case at hand, would be to appeal to the reduction of general theory to Newtonian mechanics in special cases, and to point out analogies of procedure to past Newtonian successes. J. L. Synge, in his book on general theory, briefly discusses the difficulties of justifying the use of such approximations. In general:

Approximations based on the neglect of small terms are very frequent in mathematical physics, and there is seldom any reason to object to them. One feels that if there is anything wrong, it will show up in some anomaly, and then one can revise the theory.[9]

With respect to the classical tests of general theory, Synge says:

The agonist needs no encouragement to work out, as a mathematical problem, the geodesics of space-time with the [Swartzschild] metric. The realist, on the other hand, may have some doubts. Though convinced of the validity of the geodesic hypothesis for very small bodies [e.g., that photons follow geodesics], he may wonder just what "very small" means—are the Earth and Jupiter very small? This question cannot be answered until a rational theory of the 2-body problem has been developed, and the only thing to do is to go ahead with the planetary motion and light rays.[10]

Without delving into the technicalities of the case, the point is just this: absence of experimental fit within assumed measures of goodness of fit can be construed as indicative of the inadequacy of the derivational approximations used. In such cases, the idealizations and approximations can be said to have introduced a bias in the analysis, that is, a systematic distortion. Given this bias, theoretical prediction had better *not* satisfy measures of goodness of fit based on an analysis of random experimental errors. The existence of derivational bias is the fundamental reason why statistical theories cannot provide a complete account of the concept of adequacy of experimental fit.

A simple example from freshman physics should drive home the point. If we were to ignore the rotational inertia of the suspension string or rod of a pendulum, then we would introduce a systematic bias in our prediction. If this inertia is large enough, then a statistical analysis of random experimental errors cannot help. This was one of the lessons to be learned from laboratory class: if data were to be fudged, they had to *exceed* estimated experimental error and be displaced in a way that corresponded to the bias introduced by the analysis.

Assuming, then, the insidiousness of bias owing to idealization, our original problem remains: how are idealizations justified? An obvious answer is that satisfaction of statistical measures of goodness of fit justifies the *combination* of idealized description and fundamental physical law.[11] Similarly, nonsatisfaction casts doubt on the combination. A Duhemian generalization would be: what is confirmed is the combination of law and counterfactual idealization; similarly, it is only the combination that is disconfirmed. This is a cheerful result for the antirealist, since it restricts confirmation to what is, strictly speaking, false, that is, to idealizations and approximations.[12] If, however, the history of science is to be a normative guide, then this generalization is to be resisted. This is because in several key episodes in the history of science, fundamental law (or the theoretical basis of the calculation) was held to be confirmed even though the combination of law and idealization yielded predictions that *exceeded* estimated experimental errors. Similarly, there was disconfirmation of law (or theoretical basis) *simpliciter* when goodness of fit was not satisfied. My claim is this: *sometimes* it is possible to argue that failure of fit is due to the bias introduced by idealization, and *sometimes* it is possible to argue that it is not due to such bias. In the former case, there is confirmation *even though* estimated experimental errors are exceeded. In the latter case, there is disconfirmation of law (or theoretical basis).

A particularly clear example is provided by kinetic theory and specific heats. On the basis of spectral lines and other phenomena, it was known that gas molecules have internal structure. Adding structure to the billiard ball molecules of kinetic theory, then, should have afforded some improvement in the predictions about specific heats. However, it was formally provable that adding degrees of freedom of motion (a consequence of internal structure) would cause a divergence from experimental values. Treating the molecules more realistically would make the predictions about specific heats worse, not better. Hence, there was a disconfirmation of kinetic theory; it was not possible to improve experimental fit; the bias was not eliminable. A similar example is provided by attempts to show that more realistic accounts of the Michelson-Morley experiment would accommodate the null result. Michelson's single-ray account is really not an account at all, since it does not entail a finitely-sized shifting fringe pattern. H. A. Lorentz was able to give a generalized proof, based on C. Huygens's principle and the calculus of variations, showing that the null prediction would remain despite improvements in the realism of the description of experimental conditions.

An example of eliminable bias is provided by Newton's response to critics of his *experimentum crucis*. As they noted, color separation and dispersion *are* observable after the second prism. Newton answered by arguing for this conditional: if the finite size of the apertures is taken into account (and it was not in his published treatment), then an improved prediction will result, one that allows for some color separation and image dispersion. Newton did not actually construct this improved account; he merely argued that it could be done. There are many reasons that block actual constructions of improved accounts: absence of necessary auxiliary theory, absence of required data, absence or impossibility of analytic techniques. Given these reasons, one finds arguments of the sort: *if* these difficulties were overcome, then improved experimental fit would be possible. R. S. Shankland's analysis of D. C. Miller's positive Michelson-Morley results provides an example of such argumentation.[13] If adequate theories of expansion and contraction due to thermal disturbances were available, then Special Theory could be made to yield predictions closer to the actual systematic shifts observed.

In compressed slogan form, the view proposed is this: a theory is *confirmed* if it can be shown that it is possible to show that more accurate but still idealized or approximate descriptions will lead to improved experimental fit; a theory is *disconfirmed* when it can be shown that such improvement is impossible. Since I have argued for and clarified this account elsewhere, I shall say no more in its support.[14] Assuming its correctness, we can now answer the question of what justifies idealized descriptions. In one sense, I want to deny the meaningfulness of the question. Our interest, so far, has been the confirmation or justification of theories. Our problem was that the predictions of theories and idealizations rarely, if ever, are compatible with estimated experimental error. To achieve confirmation, the blame for this failure of experimental fit must be placed on the idealized initial conditions. That is, the idealizations are *not* justified. On the other hand, theory must be logically attachable to descriptions of the world in a way that makes predictions computationally possible in fact and not just in principle. Otherwise, the process of confirmation and disconfirmation cannot even get started. Idealizations can be said, then, to be prima facie justified when they allow for practical computability. *If* it can be shown that more realistic initial conditions will lead via theory to correspondingly more accurate predictions, *then* the original, highly idealized initial conditions *are* justified in the sense that they provided the starting point for a successful confirmational process. And remember that arguing that improvement is possible need not entail actually constructing such accounts. (Similarly, although somewhat paradoxically, an idealization is justified if it provides the starting point for a successful disconfirmation.) This analysis of justified idealization also explains what it means to be *approximately true*. To say that an idealization is approximately true means that it is justified as above.

I shall now consider the consequences of the above sort of *converging counterfactual* theory of confirmation for the issue of scientific realism. Realist and antirealist perhaps can agree on this methodological point: proceed as if one were developing ever more accurate descriptions of an existing reality. Given this agreement, an argument for realism is that cases of successful convergence to better experimental fit are *miraculous coincidences* for the antirealist. Such an argument obviously will be a close cousin of similar arguments by Smart,[15] Boyd,[16] and Putnam.[17] There is this difference: whereas they focus on the apparently miraculous nature of scientific reduction and, more generally, similarity across scientific change, my version stays fixed within the context of a single theoretical program. This restriction makes it somewhat easier to state exactly what it is that is to be explained. At least two aspects of scientific practice within a research tradition require explanation. First, there is the historical fact that increasingly more accurate and complete descriptions of the apparent objects of reference typically result in more accurate predictions. Second, there is this fact about the scientific community: convergence to experimental values given more realistic treatments is seen as confirming the theoretical basis of these calculations.

Consider Newton's famous moon test. Proposition IV of Book III of *Principia* demonstrates this counterfactual: *If* (a) the earth is stationary, (b) the earth-to-moon distance is 60 earth radii, and (c) the moon's period is $27^d7^h43^m$, then assuming inverse square gravitational attraction, bodies should fall at the surface of the earth at 15.5 feet per second squared. Bodies do not fall at this rate. But neither is the earth stationary, and the earth-to-moon distance is not 60 earth radii. These values were known to be outside the range of reasonable experimental error. Historically, Newton went on to show, in various ways, that if the counterfactuality of the antecedent were relieved, then the consequence would approach a value that was, as shown by experiment, *correspondingly* more accurate.[18] What explains the success of Newton's modifications is the simple observation that since the earth and moon are existing objects, then, assuming the correctness of Newtonian mechanics, more accurate input should lead to more accurate output. This explanation would seem agreeable even to the antirealist. Such an explanation, however, is rejected in cases where the objects whose descriptions are apparently improved are nonobservables. The antirealist response here is that it makes no sense to speak of improving the description of nonexistent objects. All that exists is a heuristic procedure which results in increasingly more accurate predictions. To describe increasing the realism of our treatment of, say, molecules as just a heuristic, is not an explanation but merely a transformation of the explanandum. What requires explanation are the success of a procedure and the very possibility of such success. That the kinetic theory of Chapman and Enskog has greater empirical adequacy than Jeans's simple mean free path approach is uncontroversial fact. The question is why. By denying the notion of increasingly more accurate descriptions of nonobservables, the antirealist shuts himself off from the most reasonable explanation.

The antirealist has a powerful answer to this sort of proposed objection. There were, roughly speaking, increasingly more realistic theories of phlogiston. (And even if this is not historically as accurate as we would like, we can easily imagine it to have been so.) But there is no phlogiston; hence, there can be no account of this historical improvement in terms of it. There are several realist responses to be made here. First, before theory replacement, the existence of phlogiston is, or so the realist would claim, the best explanation of improvement. Note that it is not a requirement of realism that our current theories be true, only that we should make this our aim. And believing a theory is to believe it to be true. After theory replacement, one explains the success of phlogiston theory in terms of oxygen theory. Obviously, this cannot be done simply by mapping 'oxygen' into 'phlogiston'. These sorts of explanations, before and after theory replacement, are denied antirealists.

I shall now turn to the second aspect of progress within a tradition that requires explanation: the *confirmational cogency* of improvability of experimental fit by means of more realistic idealizations. Consider again Newton's moon test. He was able to show that if more accurate data were used in the descriptions of earth and moon, then more accurate predictions would result. This is seen as confirmation of Newton's laws. The question is: what justifies such an appraisal? For both realist and antirealist, the answer is plain. To be a physical object is to be susceptible to increasingly more accurate and correspondingly more adequate analyses on the basis of correct theories. Since the earth and moon are physical objects, and since they have been shown susceptible to improvable analysis on the basis of Newton's laws, it follows that Newton's laws are confirmed or have received some confirmation. But the antirealist will deny this explanation of confirmatory value when the objects in question are nonobservable. The antirealist, therefore, must either bite the bullet and deny that explanation is required in such cases, or provide an antirealist account of the confirmational cogency of such cases.

Say the antirealist were to adopt the line of simply *defining* confirmation to be the following: a theory is confirmed (i.e., is to be accepted as empirically adequate or whatever) when it can be shown that increasing the realism of the initial conditions (where this is judged according to scientific standards) leads to increasingly more accurate predictions. The definition is *normative*: this is how we should view theory acceptance. And the argument for its acceptance is that it reflects scientific practice (i.e., assume that my examples are typical of good scientific practice). We will let surface grammar be our guide for confirmation, but not for truth. There is an interesting analogy here with causal and necessary-and-sufficient condition theories of natural kinds. The advantage of a causal theory is that it automatically remains consistent with current scientific theory: x is gold if it is *relevantly like* this sample. If there were a final science, then this advantage would cease and the two linguistic theories would collapse into one. Here is the analogy. The above antirealist definition of confirmation will always be one step behind current scientific practice.

Consider the following situation. Because of observational and analytic shortcomings, we cannot make our idealized inputs to some theory more realistic and retain computability. However, we can vary them and retain practical computability, but not in ways where the overall realism is comparable or converging. If, in such a case, the predictions exhibit a randomlike scatter around experimental values, then we ought to accept the theory as confirmed. This is because a correct theory of observable and unobservable entities will yield, on the basis of descriptions of initial conditions "randomly" distributed about the true, predictions randomly distributed around experimental values. The predictions in such cases are said to be *robust* with respect to input. An actual instance of this confirmational methodology is given by W. Wimsatt:[19] M. J. Wade's experimental test, using laboratory populations of the floor beetle *Tribolium*, of the relative efficacy of individual and group selection.

Now, how is the antirealist to respond to this case? He will have to modify his definition of confirmation in an ad hoc way. Similarly, those who held acids to be proton donors have to modify their definition in an ad hoc way to accommodate scientific change. Stated another way: an antirealist *substitutional semantics* for scientific practice may be possible; however, it will be ad hoc and lack the rationale of a realistic referential semantics.

Obviously, none of what I have said, by itself, decides the issue of scientific realism. My purpose was the more modest one of showing how argumentation on this issue must change in order to accommodate increasingly more realistic descriptions of the connection between data and theory. This connection, I claim, is based on layers of counterfactual reasoning, where what controls the truth values of the counterfactuals changes from layer to layer.[20] I also tried to show that confirmation of scientific law was possible even though statistical measures of goodness of fit were violated. Such confirmation occurs when it can be shown that it is *possible* to improve predictions on the basis of more accurate descriptions of the objects of analysis. And to show such possibility need not entail, as the history of science demonstrates, actually constructing better analyses with better predictions. Assuming the correctness of these basic features, it appears that explanation is required of the historical instances where improved predictions resulted from improved descriptions of the objects of analysis. Explanation is also apparently required of why such instances should be seen as confirmatory of the scientific laws used in the analyses. Finally, antirealists are denied the obvious explanatory benefits of postulation of the existence of nonobservable objects. If my proposed analysis is circular, then I claim it to be of larger and more realistic radius than traditional discussions of realism and antirealism.

NOTES

1. C. Glymour, *Theory and Evidence* (Princeton: Princeton University Press, 1980).

2. Bas C. van Fraassen, *The Scientific Image* (Oxford: Clarendon Press, 1980).

3. Van Fraassen gives this analysis of scientific realism: "Science aims to give us, in its theories, a literally true story of what the world is like; and acceptance of scientific theory involves the belief that it is true." (Ibid., 8.) Since much of what this can mean depends on a theory of canonical form, I shall rest content with this analysis. By 'realism' I will not mean what Laudan has called "convergent realism": roughly that mature theories are converging on the truth because they all are referring to the same things. See L. Laudan, "A Confutation of Convergent Realism," this volume.

4. See S. Weinberg, *Gravitation and Cosmology* (New York: John Wiley and Sons, 1972), 175-194.

5. Ibid., 191.

6. The notion of a hierarchy of structures connecting data and theory comes from P. Suppes, *Logic, Methodology and Philosophy of Science: Proceedings of the 1960 International Congress* (Stanford: Stanford University Press, 1963), 252-261. For an antirealist interpretation of the starlight bending experiment, see the

dissertation of my student S. Humphrey, "An Anti-Realist Conception of Theories in Mathematical Physics," Ohio State University, 1981.

7. Sir James Jeans, *An Introduction to the Kinetic Theory of Gases* (Cambridge: Cambridge University Press, 1940), 63.

8. Weinberg, *Gravitation and Cosmology*, 191.

9. J. L. Synge, *Relativity: The General Theory* (Amsterdam: North Holland, 1960), 57.

10. Ibid., 290.

11. Because there are so many theoretical layers in the path from theory to data, there is the possibility that mistakes or errors at one level may be fortuitously canceled by errors at other levels. I ignore this complication in what follows.

12. This is a cheerful result if one assumes a Russellian theory of reference. Since the idealizations are counterfactual, nothing per se satisfies them. Therefore, there is no question of the existence of the entities apparently referred to by scientists. Science does not aim for truth, and its theories are not to be allowed as true literal descriptions.

13. For more details on the cases discussed, and for other related examples, see my "Newton's Advertised Precision and His Refutation of the Received Laws of Refraction," in *Studies in Perception: Interrelations in History and Philosophy of Science*, ed. P. K. Machmer and R. G. Turnbull (Ohio State University Press, 1977); "Feyerabend, Brownian Motion, and the Hiddenness of Refuting Facts," *Philosophy of Science* 44: 225-247; "Newton's *Experimentum Crucis* and the Logic of Idealization and Theory Refutation," *Studies in History and Philosophy of Science* 9: 51-77; and "Newton's Demonstration of Universal Gravitation and Philosophical Theories of Confirmation," in *Minnesota Studies in the Philosophy of Science*, vol. 11, ed. J. Earman (University of Minnesota Press, 1983).

14. In my "Idealization, Explanation, and Confirmation," in *PSA 1980*, vol. 1, ed. P. Asquith and R. N. Giere (Philosophy of Science Association), 336-352. I also discuss in this paper similarities and differences between my views and those of Imre Lakatos and T. S. Kuhn. The idea of giving a counterfactual interpretation to the sort of activity discussed was suggested to me by Ron Giere. For work similar to mine, see N. Koertge, "Theory Change in Science," in *Conceptual Change*, ed. G. Pearce and R. Maynard (Dordrecht: D. Reidel, 1973), 167-198; and W. Wimsatt, "Robustness, Reliability and Multiple-Determination in Science," in *Knowing and Validating in the Social Sciences: A Tribute to Donald Campbell*, ed. M. Brewer and B. Collins Jossey-Base (San Francisco: 1981).

15. J. J. C. Smart, *Philosophy and Scientific Realism* (London: Routledge and Kegan Paul, 1963).

16. R. Boyd, "Scientific Realism and Naturalistic Epistemology," in *PSA 1980*, vol. 2, ed. P. Asquith and R. N. Giere.

17. H. Putnam, "What is Mathematical Truth?", *Mathematics, Matter and Method* (Cambridge: Cambridge University Press, 1975), 60-78.

18. For a more accurate statement of Newton's procedure and more details on his demonstration of universal gravitation, see my "Newton's Demonstration of Universal Gravitation and Philosophical Theories of Confirmation."

19. Wimsatt, *Knowing and Validating in the Social Sciences: A Tribute to Donald Campbell*.

20. This paper has concentrated on the counterfactual nature of idealizations. I have not dealt with the effect the hierarchical structure of counterfactuals has on confirmation. Obviously, it serves to complicate matters greatly. In the case of the starlight bending experiment, for example, the difficulties at the lower levels of analysis are sufficiently great that there is no pressure exerted upward to develop more realistic relativistic analyses. This explains, in part, why the experiment is not seen currently as providing much in the way of confirmational value. See H. von Klüber, "The Determination of Einstein's Light-Deflection in the Gravitational Field of the Sun," in *Vistas in Astronomy*, vol. 3, ed. A. Beer (Pergamon Press, 1960), and J. Earman and C. Glymour, "Relativity and Eclipses: The British Eclipse Expeditions of 1919 and Their Predecessors," *Historical Studies in the Physical Sciences* 11: 49-85.

6

What Kind of Explanation is Truth?

Michael Levin

I

Many philosophers call themselves scientific realists because realism appears to offer the only explanation for the predictive success of science. What reason can there be for my happy arrival at my destination, other than the literal truth of the theories that guided the construction and flight of my plane? This practical success by no means entails realism, since it is possible that airplanes fly and, yet, that the aerodynamic theory of lift—particles hitting wing surfaces, and so on—is wrong. Still, it would be hard to understand how that could be; hard, that is, to swallow what would have to be a cosmically coincidental consilience between our plans and our ideas. At least realism explains this consilience, or seems to. Conversely, realism predicts that "true" theories will work; "true" theories do work; hence realism—or so it seems.

Before turning to this issue, I should explain that I understand "scientific realism" in a somewhat old-fashioned way, the way of Nagel in *The Structure of Science:* theoretical statements, read literally, possess definitive truth values, and, if true, the objects they postulate are on a par ontologically with ordinary objects of perception.[1] I also accept the more recent characterization of realism as the view that scientific theories aim to be true or very nearly true. 'Antirealism' I take to be what used to be called "instrumentalism," the thesis that the nonobservational portions of even "true" theories are not really assertions about the world, but devices of some sort for, say, codifying relations between observable variables. With regard to antirealism, then, I depart from such writers as van Fraassen and Laudan, who base antirealism on the failure of observational data ever to entail theory.[2] So understood, antirealism becomes the claim that any theory, however well confirmed and widely accepted in scientific practice, might well be and might eventually be revealed to be false. One must agree: theory does and must transcend data. But this does not even speak to what I regard as the central issue: the semantics of the theoretical portions of science. The antirealism that interests me does not regard theories as possibly false despite all evidence; it regards them as devoid of truth value. The committed instrumentalist—Duhem, for example—wants to say that even "true" scientific theories, theories that give the best account of the world that it is possible to give, are not really *true*. The claim that any theory we think is true might not actually be true is much weaker, and ignores the relatively a priori dimension of the instrumentalist's critique of discourse about unobservables. The instrumentalist thinks that talk about unobservables *cannot possibly* be any more than a way to weave together facts about observables, and this is a far stronger and, to me, more interesting claim than the pallid truism that what we hold true might not be so. In thus endorsing the formulations of old-fashioned instrumentalism against its contemporary namesakes, I do not mean, thereby, to imply that I find the formulation clear. In fact, I do not, for reasons I will discuss below. But it is the spirit behind the formulation that affords the proper contrast to realism.

I consider myself a scientific realist, but I am ready to abandon the popular argument for realism broached in the first paragraph. What has come to bother me is grave uncertainty about the kind of explanation realism gives of the success of the successful parts of science. One normally explains a phenomenon by citing the *mechanism* that brought it about or sustains it, where the mechanism itself is understood by subsumption under causal, or at least time-dependent, laws. I am aware that many philosophers profess to discern noneventival explananda, and consequently nonmechanical explanantia. Why is there no minuet in Mozart's *Prague* Symphony? (An aesthetic explanandum.) Why is slavery wrong? (A moral explanandum.) I myself doubt this, but I will not argue the point here. It is clear, at least, that explanations of the workings and successes of artifacts are virtually always mechanistic. You explain why or how a washing machine cleans dirty clothing by exhibiting the vents admitting hot water, describing the emulsive effect of soap on grease, and concluding with an account of centrifugal force in the rinse cycle. Presumably, an explanation even of the *Prague* Symphony as a successful artifact—why it moves listeners—will involve psychological mechanisms.

Now, it is perfectly plain that whatever more they may be, theories are artifacts. They are man-made devices. There was a time when planets existed and described orbits without there being theories about them—at least, without recognized theories, should one insist that there have always been theory-inscriptions embedded in material objects. This is not to deny the applicability of semantic predicates such as 'is true' to theories, in contrast to most other artifacts, such as knives, which lack semantic properties. Nor is it to insist that theories have purposes. Perhaps theories have *no* purpose, beyond truth-telling. But they can be used for successful prediction. So here we have an artifact that is successful in a certain use. An explanation for this phenomenon must manifestly be mechanical.

And here is my problem: what kind of *mechanism* is truth? How does the truth of a theory bring about, cause or create, its issuance of successful predictions? Here, I think, we are stumped. Truth, like Mae West's goodness, has nothing to do with it. "By being true" never satisfactorily answers the question, Why did such and such belief lead to correct expectations? The answer always lies elsewhere.

Consider again the question of why airplanes stay up. Note that the realist's more complex explanandum, Why do our beliefs about air flow lead to aircraft that stay up? is just a roundabout way of asking, Why do objects of a certain design stay up?, that is, Why do airplanes stay up? Surely the reason airplanes stay up is not

(1) "The pressure on the underside of a moving airfoil is greater than the pressure on its overside" is true,

but rather

(2) The pressure on the underside of a moving airfoil *is* greater than the pressure on its overside.

In saying this, I am not trying to resuscitate the redundancy theory of truth. The familiar arguments for the need and well-definedness of the truth-predicate are compelling. I would go further and maintain against, for example, Hartry Field,[3] that the standard Tarskian definition actually does tell us "what all true statements have in common." Tarski tells us that all true conjunctions have in common the truth of each conjunct, that each true existential generalization is such that its matrix is satisfied by at least one sequence, and so on. To be sure, at the level of the basis clauses, the definition goes strongly extensional, but that is the way it ought to go. The truths "Ron Reagan is a man" and "This tulip is red" are not shown to have much in common: each is a matter of a different object (or, more strictly, sequence) satisfying a different primitive open sentence. But *do* they have more in common than being a man has in common with being red, which is to say, very little? I cannot see that they do, or at least more than Tarski gives them. In short, I see truth as a genuinely general trait of sentences that Tarski's definition captures as well as it can be captured.

Nor would I deny that truth has a very important place in certain analytical enterprises. You cannot define knowledge without it, since knowledge is true belief plus something more. In terms of the topic that concerns us, the definition of explanation also requires truth. Explanations must be true—the most fascinating story about elves in my watch, which rigorously entails the explanandum of my watch keeping time, cannot explain why my watch keeps time for the simple reason that it is untrue.

So my point right now is precisely that even in (1) the explanatory work is clearly done within the quotes. Explanation by appeal to truth in this case, as in any particular case of scientific success, works by a rapid and implicit disquotation of the truth mentioned. Airplanes stay up because of the pressure differential mentioned, not because the mention of a pressure differential is the mention of a truth or even a pertinent truth.

The explanation of the success of a theory lies within the theory itself. The theory itself explains why it is successful. Indeed, on reflection, there is no such explanandum as the "success" of a particular theory, nor any such explanandum as the success of science as a whole—science conceived as a time-indexed conjunction of successful theories. (Thus, there is no danger of self-referential paradox or ungroundedness in saying that theories "explain their own success.") A theory's successes are the true predictions it has made with its own internal resources. Conjoin them and you have its success, but you do not have any further phenomenon which the theory in question fails to explain and which may perhaps be explained by some such other hypothesis as truth. To explain a conjunction, explain its conjuncts.

Granted, this precept has exceptions, usually when the explanandum involves a selection process. The explanation of why each mother cat responds to her kittens' cries doubtless involves the release of certain hormones in each mother cat by tones of a certain frequency. And, given the finitude of felines, "All mother cats respond to baby kittens crying through release of hormone so-and-so" is equivalent to the finite conjunction of its positive instances. Yet we are left in the dark about why there are not any mother cats who do not respond to their babies, and light can be shed on this only by a story about the evolutionary sieve that selected out indifferent mother cats. But this exception to the precept about conjunctions is no help to the realist, since the truth of a theory is no more likely a mechanism for sifting out false predictions the theory might have made than it is likely as the mechanism by which the theory made true predictions.

I should note, for the sake of completeness, that falsity is no better an explanation of failure than truth is of success. Phlogiston theory failed to predict the partial vacuum created by combustion, not because it was false, but because burning objects combine with atmospheric oxygen, while phlogiston theory says that burning objects give something off. Of course, burning objects do not give anything off, and that makes phlogiston theory false; but the particular account it gave of burning is why it broke down, if there must be a why of it at all.

Now, one is liable to object that I have overlooked what I previously conceded, that there can be no explanation without truth. But to concede this is not to find explanatory material in the truth feature of an explanation. The truth requirement for explanation is better expressed by saying that true statements that entail statements describing phenomena are *explanations* because they are true, than by saying that explanations explain things because they are true. In concrete cases, we always trade in the general remark "that cannot be the explanation because it is not true" for the explicit negation of the proffered explanation-candidate. The best insight into the failure of the elf-explanation is afforded by the rebuttal, "an elf in my watch cannot be why my watch keeps time, since there *is not* any elf in my watch."

In reply, the champion of truth might want to know how I distinguish explanation-candidates that fail on grounds of falsehood from those that fail on such other grounds as circularity. If someone is told that his watch keeps time because there is a miniature watch inside it, he will naturally frown and ask how the miniature watch works: "First you tell me about elves that are not there and then you ask me to assume the very thing I do not understand." The contrast between explanations that fail in point of truth and those that fail in point of intelligibility can be made without anyone supposing that truth is part of what does the explaining in a successful explanation.

Truth, then, is an unsuitable explanation for success. The principal argument for realism, indeed one popular formulation of realism itself, has come apart. Instead of retreating to instrumentalism, however, I persist in realism. I persist in distinguishing the true parts of a particular theory from its mere calculative devices, and I even persist in maintaining that this distinction sheds light on the explanatory power of particular theories. But again, this distinction can be made only within a given theory itself and, again, the explanation of the success of the true, as opposed to merely calculative, parts of a theory consists simply in letting the theory speak for itself.

To give substance to these remarks, consider a concrete case, the special theory of relativity (STR). In STR, the wave-front of a beam of light moves at the same velocity c in the reference-frames F and F' even if F and F' are moving relative to each other. It is this assumption that explains the Michelson-Morley observations, and that predicts, in an explanatory manner, certain characteristics of the Doppler effect and mass changes in accelerated particles. Let us say that this assumption belongs to the *real* part of STR. In contrast consider Minkowski's graphical representation of STR. When position is graphed against time, motionlessness becomes a vertical "world line," motion at constant velocity is a sloping straight line, accelerated motion is a curved line; all vectors fall within the line representing c. This way of representing STR is lucid and mathematically convenient, for it expresses STR in terms of simple and antecedently familiar geometrical forms. Thus, time dilation can be expressed by saying that the calibration curve for pairs of reference frames is a hyperbola, as opposed to the circle it works out to be under classical assumptions. Let us say Minkowski geometry belongs to the *calculative* part of STR.

Now, it is perfectly evident that some features of STR in Minkowski's version are artifacts of the symbolism. No one wonders why moving objects do not tip over. You cannot explain Pavarotti's mass increment during a trip to La Scala by a thickening in his world line. There is, perhaps, a question about whether light hypercones belong to the real or calculative part of STR.

There is thus a genuine distinction between the real part of STR and its calculative part. Moreover, lack of clarity about this distinction necessarily obscures the explanatory potential of STR. I even grant that marking this distinction is a philosopher's job, and that a physicist who marks it has put on his philosopher's hat. I draw the line by pointing out that the real/calculative distinction for STR was made relative to STR, and that there is no evident way to apply this distinction to theories generally. The real/calculative distinction makes sense only locally, only within a given theory.

I stress this point because the realist/nonrealist debate makes sense only if the real/calculative distinction can be given global sense. The instrumentalist says: let us take the calculative perspective toward even the ostensibly and self-styled "real" parts of particular theories. The realist says: let us not. What remains unclear to me is that there is a global calculative perspective to take. I understand someone who advises construing the constancy of the velocity of light as part of the notation of STR. At least, I understand him well enough to accuse him of not understanding the theory of relativity. But someone who says "let's take to the underpressure on an airfoil the same attitude we take to Minkowski slope in STR" has left me behind. The attitude we take toward Minkowski slope in STR is that it is a way of representing motion in space-time. Are we then to say that underpressure is or is not a way of representing motion in space-time? The instrumentalist does no better by counseling that we transfer our attitude toward Minkowski slope to *talk* about overpressure. After all, talk about overpressure is not a way of representing motion in space-time (at least not in the relevant sense).

"All parts of all theories are calculative devices" is, so far as I can tell, empty sound, just as "the realistic parts of all theories are more than calculative devices" is a useless tautology. Anyone who claims the former must explain *what* is being calculated. I suspect we have here a classic case of philosophers gripped by a picture, taking discourse appropriate in a limited context and overextending it. They take what philosopher/scientists do when contrasting the real and calculative parts of a specific theory, and suppose that this same contrast can be made when the topic is all theories, or two alternative metaphysical theories about theories. I do not see that this extension is possible. Certainly, lack of clarity about where some portion of a specific theory belongs—are light hypercones in 4-space real or calculative?—does not support a global real/calculative distinction. It suggests, at most, that the collective scientific mind is itself unclear about the nature of that part of the world the theory in question treats. Similarly, the current impasse in quantum mechanics, commonly exhibited as a refutation of realism, may instead be taken to indicate areas in which our understanding is tentative and neither a realist nor an instrumentalist view can yet be confidently advanced.

I grant that certain transtheoretical intuitions play, or seem to play, a general role in adjudicating real/calculative boundaries. Take the question of whether space-time Pythagorean distance in special relativity is a real physical magnitude or simply a way of rephrasing the constancy of the term $\Delta x^2 - c^2 \Delta t^2$. Those who take the former view will probably appeal to invariance as a general criterion of physical reality. It is the same criterion used by social scientists who require that a test of a mental ability be validated—that is, that subjects' scores on it are relatively constant across lifetimes. Only then will they assume that the test tests some independent, underlying reality. In contrast, cosmologists evidently do not take invariance as definitive, or else they would dismiss geometrodynamical theories in which the geometry of space-time is variable. By the same token, a test that assigned everyone the same height (say, the apparent amount of arc subtended at 400 yards, by the dead reckoning of a myopic Indian scout) would readily be dismissed as defective, the quantity measured being likewise a test artifact. So, while cross-theoretical methods for resolving border disputes do exist, I cannot see that they can be parlayed into a general sense for the real/calculative distinction.

There is, indeed, a perfectly general criterion that scientists and philosophers seem to use throughout all of science, which might roughly be phrased as the precept: notational changes are merely calculative. More precisely, let f be some proposed change in a theory T. Then, if, for every claim S of T, there is a counterpart $f(S)$ in the changed theory that can be determined by a relatively trivial algorithm, then f is a notational change. An example is the change from Cartesian to polar coordinates. But, this criterion is absolutely no help in getting the real/calculative distinction aloft, since it *presupposes* that we can identify the factual parts of T already. To call a change of coordinates merely calculative presupposes that such items as equations of motion are already known to be part of the substance of the theory being represented. If this makes it seem strange that philosophers should have lavished so much attention on the effort to mount a global real/calculative distinction, so be it. That the only general, cross-theoretical criterion for "calculative device"—the one that seems to correspond to the intuition underlying antirealism—should presuppose that we have already identified the real parts of theories, theory by theory, is just the way things should fall out.

Virtually the same point can be put in terms of the explanatory power of theories. If we ask a physicist why STR works, he will naturally fall back on the realistic part of STR. Perhaps if he is asked the general question, "Why do some theories work?" he may say "because of their realistic parts." But in any concrete case of success, he will simply *give* the derivation of the success from the realistic part of the theory in question, duly relegating the calculative part to the role of calculation.

A final point: it is commonly remarked that the relation of explanation to prediction is asymmetrical, since there are predictions that are not potential explanations. Consider the prediction that a subject whose sentences have conformed to a set of grammatical rules will continue to conform to those rules. Without a further mechanical theory behind this conformity, "curve-fitting" grammatical theory can be predictively successful but unexplanatory. Does this require a role for truth in at least a nonexplanatory, predictive theory of success? But to use the hypothesis that a theory is true to predict that the theory will continue to churn out correct predictions, on the basis of its past success, is simply to use the past success of the theory as evidence for the theory itself and thus for its predictions. Appeal to the truth of the theory has once again collapsed into talk of the theory itself and the evidence for it. When saddled with a predictively successful but unexplanatory theory, we generally search for a mechanism that accounts for both the phenomena the theory predicted and the predictive success of the theory itself. Even if the assumption of truth were somehow construable as that which warranted continued use of a theory for purposes of prediction, we expect it eventually to yield to a deeper mechanistic theory which preserves all the predictive power of what it replaces.

II

It is tempting for the realist to retreat to the view that the ontological commitments—the objectual posits and sortal predicates said to be satisfied—of sophisticated theories are to be taken seriously, and that the success of successful theories is best explained by taking their ontological commitments seriously. On this view, the success of, for example, particle physics would be a miracle unless we suppose that the objects talked about by particle physics *really exist.* Of course, there is the complication that particle physics may be a little off, and that, for example, hadrons have a different internal structure than they are accorded in contemporary hadron-theory. We might have a relatively successful theory whose literal ontological commitments are wrong and, therefore, whose success cannot be attributed to the reality of its posits. As a consequence, this new form of realism will be further weakened to say that successful theories are successful because there are things like, or very nearly like, what such theories posit, and that the terms of such theories *refer* to objects that the theory misdescribes. The matter can then be clarified by enlisting some version of the speaker's reference/semantic reference distinction— so that users of the theory can be said to refer to things that do not satisfy the descriptive terms they are using—or by moving to Ramsey sentences, as W. V. O. Quine and Neil Wilson suggest.

But all this imputes to or imposes on scientific theories too much of the ontological idiom of logicians. To say that hadron theory correctly predicts the results of accelerator experiments because 'hadron' refers or because (Ex) ('hadron' refers to x) is simply to say that hadron theory works because there are hadrons as described, more or less, by the theory—and this, again, is simply to *reiterate the theory,* to say there are hadrons. Even in this new sense, it is not realism that explains the success of particle theory, but particle theory itself. Once again, we get all the explanation of a theory's success we will ever need if we only let the theory speak for itself.

Philosophers sometimes think that scientific theories cannot speak for themselves very well. They think theories stutter. They think, in particular, that there is an intrinsic lack of clarity about what a scientific theory says exists that can be resolved only by (i) articulating a criterion of ontological commitment, then (ii) regimenting theories, so that (iii) we all may see by the lights of (i) what the result of (ii) says there is. This seems to me a complete mistake.[4] Theories say what they think there is in the best and only way that theories can: by using such idioms as "there are," "there exist," and the like. After all—and I think S. Kripke has made a similar point—when we introduce such devices for ontological regimentation as the existential quantifier and one or many sorts of variables, we have to say what those notations mean, and we have to do so in previously comprehensible natural language if they are to be more than symbols for formal play. We have to explain what "(Ex)" means if we are to use it to express existence or anything else, and if we use it to express existence, we must explain it in terms of the idioms of existence that scientific theories use in the first place. It is by prior mastery of these idioms that we know what the logician is driving at, and by which the logician himself has anything to drive at. These idioms, then, are the primary devices for carrying ontic implications, and we can read the ontological implications directly from a theory at least as well as we can from its first-order regimentation. Once again, we do best by letting theories speak for themselves.[5]

Not merely the idea of "notational device" but further that of "truthlikeness" must be relativized to particular theories. Popper measures the truthlikeness of a theory in terms of the number of its true consequences: theory T is nearer the truth than comparable theory T' if T has more true consequences than T', but not more false consequences. Coordinate with the technical difficulties this definition faces is a certain wrongheadedness in spirit. It seems to me that verisimilitude has meaning only relative to a specific parameter, $p(x)$. A theory T is close to the truth with respect of p, as the values T ascribes to p are close to the real values p takes. If theories T and T' share a parameter p, then T is closer to the truth than T' with respect to p if T always assigns values to p closer to its actual values than does T'. Without going into all the technical variants that can be played on such a conception—the weighting of parameters to get a global measure of truthlikeness for a particular theory or a more accurate comparison between theories, the kinds of partial orderings that can be induced by comparing pairs of the form (T,p)—it is clear that this notion differs from Popper's in significant ways. It covers cases in which two theories constantly err on the value of a parameter, but one is wrong by less than the other. Two theories can be comparable with respect of one parameter but not another—a case in which, under Popper's conception, the theories are not comparable at all. Also, on my view, closeness to truth for a theory makes sense *only* when we are talking about an additive, ideally real-valued parameter. Take the correct value of a parameter. Now take the value(s) of that parameter assigned by a theory. The neighborhood of the actual value within which the assigned value(s) fall is "how close" the theory comes to truth. Not only is truthlikeness thus distinct from totality of true predictions in any larger sense, but the truthlikeness of a theory becomes one more fact internal to the theory and as capable of exact expression within the language of the theory as the facts the theory itself expresses. Naturally, trapped as we are within our own best theory, which may not be true, we may not know until too late just how close to true, with respect of a given parameter, our theory is. But how close to true it actually is is expressible in the vocabulary of the theory itself, even if we never express it.

Why, then, be a scientific realist, if belief in the truth of science is no help in understanding its success, and if belief in the referential status of theoretical terms consists simply in taking theories at face value? Well, since typical scientific theories are themselves as realistic as bodies of discourse can be, I am a scientific realist precisely because I believe science. Scientific theories themselves talk about objects in the most committal vocabulary there is, and I believe this talk because I take the evidence for scientific theories themselves seriously. Like Hume's Cleanthes, I proportion my assent to the precise degree of evidence that occurs, and the evidence that accrues to particular theories currently advanced by scientists is good indeed.

If scientific realism is simply believing scientific theories, then various forms of instrumentalism must be, in one way or another, simply dissent from science. And I think this is the way it turns out. I cannot document this large claim here, but a good example is Ernst Mach's segue from quasi-phenomenalism as a philosophical thesis to dogmatic disbelief in atoms. Another example is the treatment of intelligence by obscurantists such as Stephen Jay Gould who maintain IQ tests measure only "the ability to do well on IQ tests." This begins as instrumentalism about mental testing, and issues in a denial of any underlying reality that explains performance on IQ tests and the high correlation between IQ and other testable skills. Thus, Gould ends up repudiating or ignoring facts about these correlations themselves.[6] He fails to see that the question of the heritability of multiple basic abilities arises in just the same way as the heritability of g does. And he ignores studies in which deprived minorities do very well on IQ tests normed on a white population—a refutation, at the level of observables, of the idea that IQ tests only test a narrow-spectrum ability acquired by propitious training.

The only interesting ground for general skepticism about science is what I have called the "inductive argument against induction": we have failed so often we are likely to be failing right now. Our past failures are trustworthy evidence that what we take to be evidence is not to be trusted.[7] Rather than repeat my own animadversions against this argument, I will repeat Cleanthes' great rhetorical question to the skeptical Philo in Hume's *Dialogues*: "And what would you say to one who, having nothing particular to object to the arguments of Copernicus and Galileo for the motion of the earth, should withhold his assent on that general principle that these subjects were too magnificent and remote to be explained by the narrow and fallacious reason of mankind?" "Nothing particular to object"; the failure of Newton to get it right is no argument against Einstein's theory, since Einstein's theory does not entail that Newton got it right. Evidence against Einstein's theory can only come from astronomical observation, from observation of what Einstein's theory *is* about. Should observation be at variance with prediction, then we will have "something particular to object."

By throwing my lot in with science, I do not thereby relativize truth or existence to science, or to science at any particular time. Obviously, what I *believe* to be true is dictated by what I believe. In fact, it *is* what I believe. But it hardly follows that the notion of truth itself has to be relativized to "conceptual frameworks," just because I cannot leave my own to make judgments of truth. Similarly, what I take to exist, and what I take myself to refer to, are dictated by my current beliefs. What else should I go on? But this hardly entails that reference and existence are notions incapable of absolute definition.[8]

This comment on the prospects for an absolute notion of reference may seem at odds with the other, distinctly Quinean, strands of this paper. I am, in a sense, relativizing everything to science, abandoning any exterior vantage point from which philosophy can say anything intelligible about science and, in particular, dismissing the realist/instrumentalist debate as meaningless. Must I not then also adopt the naturalist epistemology that goes along with it? To be sure, but this naturalistic epistemology does not lead to Quine's notorious "inscrutability of reference," which would relativize reference and ultimately ontology to particular conceptual schemes. One can accept practically everything Quine says about language, including his behaviorism and his naturalism, and reject the inscrutability of reference. What emerges is a robust realism for the dedicated realist.

The argument for inscrutability in Quine's *Word and Object* (section 9) is quite straightforward, and the fallacy it rests on is easily identified. The heart of Quine's claim that reference is not uniquely determinable is that two stimulus-synonymous terms need not refer to the same things, and that stimulus-synonymy exhausts the objective facts about language use. Quine seems right on the first point. Any stimulation that a rabbit gives rise to is also given rise to by a temporal slice of a rabbit. Therefore, the noncoreferential terms 'rabbit' and 'rabbit slice' preserve stimulus-meaning as translations of 'Gavagai'. But why assume that the only objective fact about a speaker's use of a word is its *stimulus*-meaning? Surely, not in order to be a behaviorist or an antimentalist, since a behaviorist could as well describe the speaker's behavior by saying that he emitted "Gavagai" when a rabbit was present. Quine's argument for taking *stimulus*-meaning as what is objective about word usage is disarmingly simple, insufficiently attended to, and a superaddition to naturalism:

It is important to think of what prompts the native's assent to "Gavagai?" as stimulations and not rabbits. Stimulation can remain the same though the rabbit be supplanted by a counterfeit. Conversely, stimulation can vary in its power to prompt assent to "Gavagai?" because of variations in angle, lighting, and color contrast, though the rabbit remain the same.

The argument, then, is this: suppose that what provoked "gavagai"—what in fact provided the prompting stimulus—was in fact a whole rabbit. Had an indiscriminable decoy or malign spirit provided the same stimulus, the subject would have reacted the same way. What is doing the causal work—what supports counterfactuals, so to speak—is the stimulus, not its distal cause. The rabbit drops out as irrelevant.

This does not show that it matters not what *in fact* produced a stimulus which would have prompted the same response had it been produced differently. Our intuitions about references and causation lead us to describe a situation in which a rabbit is behind the utterance of "gavagai" as one in which a *rabbit* prompts "gavagai," not the element of stimulation that is common to that situation and a different, hypothetical one. Granted that the proximate cause of a speaker's behavior is an event that occurs at the speaker's surface,[9] once we avoid the phenomenalist error of supposing that terms refer to the proximate stimuli Quine concentrates on, why not allow that the objects of reference are the *actual* causes of these proximate causes of verbal behavior? Let the stimuli be the cause, the object of reference the actual, though not causally necessary, cause of the cause. Of the counterfactual situations Quine has us consider, it is consistent and intuitively plausible to say: "If the stimulus that provoked 'gavagai' had been caused by a clever decoy, the native would have been referring to the decoy by 'gavagai'. But it was not, so he was not." Indeed, if the stimuli usually caused by rabbits were usually caused by decoys, the natural thing to say would be that 'gavagai' refers to decoys. How odd it is to say that speakers in our world do not differ objectively with regard to the reference of 'rabbit' from speakers in a world full of clever wooden decoys! One thinks here of Putnam's twin-earth case;[10] surely, $Oscar_1$ is referring to water, and $Oscar_2$ is referring to XYZ. These are objective facts, not the indulgence of a charitable translator who translates 'water$_2$' as XYZ by the *convention* that you assign reference to what is nearest.

So I see no reason to take the stimulus-meaning of a word as exhausting what is objective about its use, and take Quine's refusal to go beyond stimulus-meaning as attributable to a simple, if unnamed, causal fallacy. If I shoot you through the heart, I kill you. That the fatal bullet or one just like it might have reached your heart with identical results from another source does not disqualify my shooting as the actual cause of your death. A flowerpot falls from a roof and dents the sidewalk. That the sidewalk would have responded identically to any other impressed force of identical magnitude and direction does not mean that we have to snap the causal chain before the flowerpot. I suggest that there is no good reason, practical or theoretical, to snap the causal chain of speech provocation where Quine does, and certainly not on the basis of his counterfactual argument.

Prompting stimulation is a big part of the objective truth about language use, and any stimulation that can be traced back to a rabbit can indeed be traced back to a rabbit-stage, though not of course to a decoy. Yet, my argument suggests that stimulus-meaning should be de-emphasized and, hence, that the inscrutability of reference up to stimulus-meaning does not really imply the inscrutability of reference absolutely. This leaves me with an obligation to specify what more there is to the objective side of language. In part, I discharge this obligation simply by remarking on how arbitrary and ungrounded is Quine's insistence that stimulus-meaning is exhaustive. More expansively, I suggest that, for all the problems surrounding mentalism, the Gricean analysis of meaning leads us back into the mind, to the intentions of speakers and, presumably, to how they intend to divide up the world. I further suggest that the conundrum about rabbits versus rabbit-stages has many of the features of such standard philosophical riddles as the five-minute-old hypothesis. We have two theories that clearly embody different pictures of the world, but that are, by design, observationally equivalent. This pushes us in the direction of saying that the difference between them is conventional— that it is six of one if we chop the world into rabbits, and a half dozen of the other if we chop it into stages. But this will not do, since there is a *clear* empirical difference between saying that speaker *s* is referring to stages and saying that *s* is referring to whole rabbits. Those two theories yield extensionally inequivalent translations of his words.

What I stress is that there is a real difference between the situation in which we are trying to construe the words of a speaker *s* who is unaware of the various possible ontologies, and the words of a speaker who has come to consciousness of them. In one case—s might be a brutal fifteenth-century peasant hunting rabbits on his lord's estate—it really does seem arbitrary to say that by "rabbit" he means something that lasts a minute or the lifetime of the rabbit. He certainly has never given the matter any thought; all that is objectively true is that if his retinae are suitably irradiated and he is asked "rabbit?" he will say "yes." It is otherwise with the lord himself, who has delved into metaphysics and essence. He has become aware that his serf's words can be construed equally well in both ways, and that even his *own* words could be so construed in the past. But now he comes to a decision about which sort of thing he shall refer to in the future. He decides, say, that it will be rabbitlike things that last as long as rabbits. Then for him there is certainly an objective fact of the matter about what he is talking about.

As I understand them, Quineans at this point deny that the lord can make this distinction *even to himself*, since the proof of the verbal pudding is in the speaking. Even his own internal resolve comes to nothing if it does not change his dispositions to verbal behavior. But if the lord himself cannot make this distinction, then no one can, Quine included. If Quine were right, we should literally not know what he was talking about when he directed our attention to the *difference* between rabbits and rabbit-stages. So Quine cannot be right, even if it takes a coming-to-consciousness-of-reference, a Hegelian subreption of Quinean reason, to make this manifest.

The scientific realist wants to finish the story of reference that Quine leaves incomplete with talk about those actual physical objects in the world that prompt referential speech. In some cases, the speaker himself may be ignorant, confused, or unconcerned about them. The determinate objects of reference are matters for an objective observer to note and integrate into a complete, naturalistic theory of man as speaker and knower—with the aid of science.

NOTES

1. E. Nagel, *The Structure of Science* (New York: Harcourt, Brace, and World, 1961), 118.

2. See the contributions of Laudan and van Fraassen to this book. Their antirealism takes theoretical statements, read literally, to have truth value. Their defenses of antirealism, however, stress, respectively, the historical failure of well-confirmed theories and problems of underdetermination [ed.].

3. H. Field, "Tarski's Theory of Truth," *Journal of Philosophy* 69, 13 (1972): 347-375.

4. I myself may have been guilty of this mistake in my book *Metaphysics and the Mind-Body Problem* (Oxford: Oxford University Press, 1979).

5. This is not to say that logicians' criteria for ontological commitment are useless, or that one is no better than another. As Quine pointed out, there is nothing like first-order notation to keep other philosophers from saying things such as "There are propositions, but propositions do not really exist." (Of course, this is an intramural debate; scientists have more sense.) Also, I still think my suggestion in *Metaphysics and the Mind-Body Problem* that the ontological commitment of a speaker's discourse is best read from a scientific metalanguage for that discourse, offers a marginally better treatment of speakers locked into an ontologically profligate idiom than does Quinean transcription into canonical notation.

I would urge a similar moral about distinctions. I do not think any distinction can be made or explained by introducing formal notation for it. If the notation is to be anything more, the annotator must explain it in his mother tongue; and this means that a distinction, made by notation, must already be available in that mother tongue. Of course, good notation can help fix a slippery distinction in one's mind, as witness scope distinctions in first-order modal logic. But this is a far cry from countenancing the imposition of a distinction via formalization on an antecedently indifferent natural language.

6. S. J. Gould, *The Mismeasure of Man* (New York: Norton, 1981). For detailed criticism of this work and of Gould's views on intelligence, see M. Levin, "Hissing the Messenger," *Policy Review* (Summer 1982): 173-178.

7. "On Theory-Change and Meaning Change," *Philosophy of Science* 46, 3 (1979): 407-423.

8. Just this fallacious inference is endorsed by Richard Rorty in *Philosophy and the Mirror of Nature* (Princeton: Princeton University Press, 1980).

9. Even in saying this we have to be careful, or nothing will ever be the cause of anything. If we require the actual cause to be such that it not only produces the effect but also such that nothing could possibly have gotten between it and the effect—such that it could not possibly have occurred without the ensuing event—then the murderer's firing of the gun is not *really* the cause of death.

10. "The Meaning of 'Meaning,'" *Philosophical Papers*, vol. 2 (Cambridge: Cambridge University Press, 1975).

My wife Margarita contributed helpful comments, especially to the second half of this paper.

7

What is Realism?

Hilary Putnam

While it is undoubtedly a good thing that "ism" words have gone out of fashion in philosophy, *some* "ism" words seem remarkably resistant to being banned. One such word is 'realism'. More and more philosophers are talking about realism these days; but very little is said about what realism is.

Whatever else realists say, they typically say that they believe in a 'correspondence theory of truth'.

When they argue *for* their position, realists typically argue against some version of idealism—in our time, this would be positivism or operationalism. (This is not in itself surprising; all philosophers attempt to shift the burden of proof to their opponents. And if one's opponent has the burden of proof, to dispose of his arguments seems a sufficient defense of one's own position.) And the typical realist argument against idealism is that it makes the success of science a *miracle*. Berkeley needed God just to account for the success of beliefs about tables and chairs (and trees in the quad); but the appeal to God has gone out of fashion in philosophy, and, in any case, Berkeley's use of God is very odd from the point of view of most theists. And the modern positivist has to leave it without explanation (the realist charges) that 'electron calculi' and 'space-time calculi' and 'DNA calculi' correctly predict observable phenomena if, in reality, there are no electrons, no curved space-time, and no DNA molecules. If there are such things, then a natural explanation of the success of these theories is that they are *partially true accounts* of how they

140

behave. And a natural account of the way in which scientific theories succeed each other—say, the way in which Einstein's Relativity succeeded Newton's Universal Gravitation—is that a partially correct, partially incorrect account of a theoretical object—say, the gravitational field, or the metric structure of space-time, or both—is replaced by a *better* account of the same object or objects. But if these objects do not really exist at all, then it is a *miracle* that a theory which speaks of gravitational action at a distance successfully predicts phenomena; it is a *miracle* that a theory which speaks of curved space-time successfully predicts phenomena; and the fact that the laws of the former theory are derivable, "in the limit," from the laws of the latter theory has no methodological significance.

I am not claiming that the positivist (or whatever) has no *rejoinder* to make to this sort of argument. He has a number: reductionist theories of the *meaning* of theoretical terms, theories of explanation, and so forth. Right now, my interest is rather in the following fact: the realist's argument turns on the success of science, or, in an earlier day, the success of commonsense material-object theory. But what does the success of science have to do with the correspondence theory of truth?—or *any* theory of truth, for that matter?

That science succeeds in making many true predictions, devising better ways of controlling nature, and so forth, is an undoubted empirical fact. If realism is an *explanation* of this fact, realism must itself be an overarching scientific *hypothesis*. And realists have often embraced this idea, and proclaimed that realism *is* an empirical hypothesis.[1] But then it is left obscure what realism has to do with theory of *truth*. In the present paper, I shall try to bring out the connection between explaining the success of knowledge and the theory of truth.

THE 'CONVERGENCE' OF SCIENTIFIC KNOWLEDGE

What I am calling 'realism' is often called 'scientific realism' by its proponents. If I avoid that term here, it is because 'scientific realist', as a label, carried a certain ideological tone—a tone more than faintly reminiscent of nineteenth-century materialism or, to be blunt about it, village atheism. Indeed, if a scientific realist is one who believes, among other things, that *all* knowledge worthy of the name is part of science, then I am not a scientific realist. But scientific knowledge is certainly an impressive part of our knowledge, and its nature and significance have concerned all the great philosophers at all interested in epistemology. So it is not surprising that both realists and idealists should claim to be philosophers of science, in *two* senses of 'of'. And if I focus on scientific knowledge in what follows, it is because the discussion has focused on it, and not out of a personal commitment to scientism.

To begin with, let me say that I think there *is* something to the idea of convergence in scientific knowledge. What there is is best explained, in my opinion, in an unpublished essay by Richard Boyd.[2] Boyd points out that all that follows from standard (positivist) philosophy of science is that later theories in a science, if they are to be better than the theories they succeed, must imply many of the observation sentences of the earlier theories (especially the true observation sentences implied by the earlier theories). It does not follow that the later theories must imply the approximate truth of the theoretical laws of the earlier theories in certain circumstances—which they typically do. In fact, preserving the *mechanisms* of the earlier theory as often as possible, which is what scientists try to do (or to show that they are limiting cases of new mechanisms), is often the *hardest* way to get a theory that keeps the old observational predictions, where they were correct, and simultaneously incorporates the new observational data. That scientists try to do this—for example, preserve conservation of energy, if they can, rather than postulate violations—is a fact, and that this strategy has led to important discoveries (from the discovery of Neptune to the discovery of the positron) is also a fact.

Boyd tries to spell out realism as an overarching empirical hypothesis by means of two principles:

1. Terms in a mature science typically *refer*.
2. The laws of a theory belonging to a mature science are typically approximately *true*.

What he attempts to show in his essay is that scientists act as they do because they *believe* (1) and (2) and that their strategy works because 1 and 2 are *true*.

One of the most interesting things about this argument is that, if it is correct, the notions of truth and reference have a causal-explanatory role in epistemology. Principles 1 and 2 are premises in an *explanation* of the behavior of scientists and the success of science—and they essentially contain concepts from referential semantics. Replacing 'true' in premise 2 (of course, Boyd's argument needs many more premises than *just* 1 and 2) by some operationalist 'substitute'—for example, 'is simple and leads to true predictions'—will not preserve the explanation.

Let us pause to see why. Suppose T_1 is the received theory in some central branch of physics (physics surely counts as a mature science, if any science does), and I am a scientist trying to find a theory T_2 to replace T_1. (Perhaps I even know of areas in which T_1 leads to false predictions.) If I believe principles 1 and 2, then I know that the laws of T_1 are (probably) approximately true. So T_2 must have a certain property—the property that the laws of T_1 are "approximately true" *when we judge from the standpoint of T_2*—or T_2 will (probably) have no chance of being true. Since I want theories that are not *just* "approximately true," but theories that have a chance of being *true*, I will only consider theories, as candidates for being T_2, which have this property, that is, theories which contain the laws of T_1 as a limiting case. But this is just the feature of the scientific method we discussed. (Boyd also discusses a great many other features of the scientific method—not just this aspect of 'convergence'; but I do not need to go into these other features here.) In fine, my knowledge of the truth of principles 1 and 2 enables me to restrict the class of candidate-theories I have to consider, and thereby increases my chance of success.

Now, if all I know is that T_1 leads to (mainly) true predictions in some observational vocabulary (a notion I have criticized elsewhere),[3] then all I know about T_2 is that it should imply most of the observation sentences implied by T_1. But it does *not* follow that it must imply the truth of the laws of T_1 in some limit. There are many other ways of constructing T_2 so that it will imply the truth of most of the observation sentences of T_1, and making T_2 imply the "approximate truth" of the laws of T_1 is often the hardest way. Nor is there any reason why T_2 should have the property that we can assign *referents* to the terms of T_1 from the standpoint of T_2. Yet, it is a fact that we can assign a referent to 'gravitational field' in Newtonian theory from the standpoint of relativity theory (though not to 'ether' or 'phlogiston'), and a referent to Mendel's 'gene' from the standpoint of present-day molecular biology, and a referent to J. Dalton's 'atom' from the standpoint of quantum mechanics. These retrospective reference assignments depend on a principle that has been called the "principle of benefit of the doubt" or the "principle of charity,"[4] but not on *unreasonable* charity. Surely, the gene discussed in molecular biology is the gene (or rather 'factor') Mendel *intended* to talk about; it is certainly what he should have intended to talk about! Again, if one believes that the terms of T_1 do have referents (and one's semantic theory incorporates the principle of benefit of the doubt), then it will be a constraint on T_2, it will narrow the class of candidate-theories, that T_2 must have the property that *from its standpoint* one can assign referents to the terms of T_1. And again, if I do not use the notions of truth and reference in philosophy of science, if all I use are "global" properties of the order of "simplicity" and "leads to true predictions," then I will have no analogue of this constraint, and I will not be able to narrow the class of candidate-theories in this way.

WHAT IF THERE WERE NO CONVERGENCE IN
SCIENTIFIC KNOWLEDGE?

Let me now approach these problems from the other end, from the problem of truth. How would our notions of truth and reference be affected if we decide there is no convergence in knowledge?

This is already the situation according to someone such as Kuhn, who is skeptical about convergence and who writes (at least in *The Structure of Scientific Revolutions*) as if the same term cannot have the same referent in different paradigms (theories belonging to or generating different paradigms correspond to different 'worlds', he says), and, even more so, from P. Feyerabend's standpoint.

Let us suppose they are right, and that 'electron' in Bohr's theory (the Bohr-Rutherford theory of the early 1900s) does not refer to what we *now* call electrons. Then it doesn't refer to *anything* we recognize in present theory and, moreover, it doesn't refer to anything from the standpoint of present theory (speaking from that standpoint, the only things Bohr could have been referring to were electrons, and if he was not referring to electrons he was not referring to anything). So if we use present theory to answer the question, was Bohr referring when he used the term 'electron'? the answer has to be "no," according to Kuhn and Feyerabend. And what other theory can we use but our own present theory? (Kant's predicament, one might call this, although Quine is very fond of it too.) Kuhn talks as if each theory does refer—namely, to *its own* 'world' of entities—but that is not true according to *any* (scientific) theory.

Feyerabend arrives at his position by the following reasoning (which Kuhn does not at all agree with—any similarity in their views on cross-theoretical reference does not come from a shared analysis of science): the introducer of a scientific term, or the experts who use it, accept certain laws as virtually necessary truths about the putative referent. Feyerabend treats these laws, or the theoretical description of the referent based on these laws, as, in effect, a *definition* of the referent (in effect, an *analytic* definition). So if we ever decide that nothing fits that exact description, then we must say that there was no such thing. If nothing fits the exact Bohr-Rutherford description of an electron, then 'electron' in the sense in which Bohr-Rutherford used it does not refer. Moreover, if the theoretical description of an electron is different in two theories, then the term 'electron' has a different sense in the two theories (since it is synonymous with different descriptions—Feyerabend does not say this explicitly, but if this isn't his argument, then he has none). In general, Feyerabend concludes, such a term can have neither a shared referent nor a shared sense in different theories (the "incommensurability of theories").

This line of reasoning can be blocked by arguing (as I have and as Saul Kripke has) that scientific terms are not synonymous with descriptions. Moreover, it is an essential principle of semantic methodology that when speakers specify a referent for a term they use by a *description* and, because of mistaken factual beliefs that these speakers have, that description fails to refer, we should assume that they would accept reasonable reformulations of their description (in cases where it is clear, given our knowledge, how their description should be reformulated so as to refer, and where there is no ambiguity about how to do it in the practical context). This is, roughly, the principle of benefit of the doubt alluded to above.

To give an example: there is nothing in the world that *exactly* fits the Bohr-Rutherford description of an electron. But there are particles which *approximately* fit Bohr's description: they have the right charge, the right mass, and they are responsible for key effects which Bohr-Rutherford explained in terms of electrons; for example, electric current in a wire is flow of these particles. The principle of benefit of the doubt dictates that we treat Bohr as referring to these particles.

Incidentally, if Bohr had not been according the benefit of the doubt to his earlier (Bohr-Rutherford period) self, he would not have continued to use the term 'electron' (without even a gloss!) when he participated in the invention of (1930s) quantum mechanics.

Coming back to Kuhn, however: we can answer Kuhn by saying there *are* entities—in fact, just the entities we now call "electrons"—that behave like Bohr's electrons in many ways (one to each hydrogen atom, negative unit charge, appropriate mass, etc.). And (this is, of course, just answering Kuhn exactly as we answered Feyerabend) the principle of benefit of the doubt dictates that we should, in these circumstances, take Bohr to have been referring to what we call "electrons." We should just say we have a different theory of the *same* entities Bohr called "electrons" back then; his term did refer.

But we can only take this line because present theory does assert the existence of entities that fill many of the *roles* Bohr's electrons were supposed to fill, even if these entities have other, very strange properties, such as the complementarity of position and momentum, which Bohr-Rutherford electrons were not supposed to have. But what if we accept a theory from the standpoint of which electrons are like phlogiston?

Then we will have to say electrons do not really exist. What if this keeps happening? What if *all* the theoretical entities postulated by one generation (molecules, genes, etc., as well as electrons) invariably do not exist from the standpoint of later science? This is, of course, a form of the old skeptical "argument from error"—how do you know you are not in error now? But it is the form in which the argument from error is a *serious* worry for many people today, and not just a philosophical doubt.

One reason this is a serious worry is that eventually the following metainduction becomes overwhelmingly compelling: *just as no term used in the science of more than fifty (or whatever) years ago referred, so it will turn out that no term used now refers* (except maybe observational terms, if there are such).

It must obviously be a desideratum for the theory of reference that this metainduction be blocked; that is one justification for the principle of benefit of the doubt. But benefit of the doubt can be *unreasonable*; we do not carry it so far as to say that 'phlogiston' referred. If there is no convergence, if later scientific theories cease having earlier theories as limiting cases, if Boyd's principles 1 and 2 are clearly false from the point of view of future science, then the principle of the benefit of the doubt will always turn out to be unreasonable—there will not be a reasonable modification of the theoretical descriptions of various entities given by earlier theories that will make those descriptions refer to entities with somewhat the same roles that do exist from the standpoint of the later theory. Reference will collapse.

But what happens to the notion of truth in theoretical science if none of the descriptive terms refer? Perhaps all theoretical sentences are false; or perhaps some convention for assigning truth-values when predicates do not refer takes over. In any case, the notion of "truth-value" becomes uninteresting for sentences containing theoretical terms. As a consequence, truth will collapse too.

Now, I want to argue that the foregoing *is not* quite what would happen. But this will turn on rather subtle logical considerations.

MATHEMATICAL INTUITIONISM: AN APPLICATION TO EMPIRICAL KNOWLEDGE

On the assumption that the reader has not studied "mathematical intuitionism" (the school of mathematical philosophy developed by L. Brouwer, A. Heyting, etc.), let me mention a few facts that I will use in what follows.

A key idea of the intuitionists is to use the logical connectives in a nonclassical sense. (Of course, intuitionists do this because they regard the classical sense as inapplicable to reasoning about infinite or potentially infinite domains.) They explain this sense—that is, they explain their meanings for the logical connectives—in terms of constructive *provability* rather than (classical) truth.

Thus:

1. Asserting *p* is asserting *p is provable*. ('*p* · ⌐*p* is not provable ¬' is a contradiction for the intuitionists.)

2. '¬*p*' (¬ is the intuitionist symbol for negation) means *it is provable that a proof of* p *would imply the provability of* 1=0 (or any other patent absurdity). In other words, ¬ *p* asserts the *absurdity* of p's *provability* (and not the classical 'falsity' of *p*).

3. '*p* · *q*' means *p is provable and* q *is provable*.

4. '*p* v *q*' *means there is a proof of* p *or a proof of* q *and one can tell which*.

5. '*p* ⊃ *q*' means *there is a method which applied to any proof of* p *yields a proof of* q (*and a proof that the method does this*).

These meanings are clearly different from the classical ones. For example, *p* v ¬ *p* (which asserts the decidability of every proposition) is not a theorem of intuitionist propositional calculus.

Now, let us reinterpret the classical connectives as follows:

1. ~ is identical with ¬ .

2. · (classical) is identified with · (intuitionist).

3. *p* v *q* (classical) is identified with ¬(¬*p* · ¬*q*).

4. *p* ⊃ *q* (classical) is identified with ¬(*p* · ¬*q*).

Then, with this interpretation, the theorems of classical propositional calculus become theorems of intuitionist propositional calculus![5] In other words, this is a translation of classical propositional calculus into intuitionist propositional calculus—not, of course, in the sense of giving the classical "meanings" of the connectives in terms of intuitionist notions but in the sense of giving the classical theorems. (This is not the only such translation, by the way.) The meanings are still not classical, if the classical connectives are reinterpreted in this way, because these meanings are explained in terms of provability and not in terms of truth and falsity.

To illustrate: classically *p* v ~ *p* asserts that every proposition is true or is false. Under the above 'conjunction-negation translation' into intuitionist logic, *p* v ~ *p* asserts ¬(¬*p* · ¬¬*p*), which says that it is absurd that a proposition and its negation are both absurd—nothing about being true or false!

One can extend all this to the quantifiers—I omit details.

This shows that contrary to what a number of philosophers (recently including Hacking) have asserted, such inference rules as *p* · *q*⊢ *p*; *p* · *q*⊢ *q*; *p*⊢ *p*v*q*; *q*⊢ *p*v*q*; ~ *p*, ~ *q*⊢ ~ (*p*v*q*) do not fix the 'meanings' of the logical connectives. Someone could accept all of these rules (and all classical tautologies, as well) and still be using the logical connectives in the nonclassical sense just described—a sense that is not truth-functional.

Suppose, now, we apply this interpretation of the logical connectives (the interpretation given by the 'conjunction-negation translation' above) to empirical science (this idea was suggested to me by reading Dummett on Truth, although he should not be held responsible for it) in the following way: replace constructive provability (in the sense of intuitionist mathematics) by constructive provability from (some suitable consistent reconstruction of) the postulates of the empirical science accepted at the time (or, if one wishes to be a realist about observation statements, those together with the set of true observation statements).[6] If the empirical science accepted at the time is itself inconsistent with the set of true observation statements—because it implies a false prediction—then some appropriate subset would have to be specified, but I shall not consider here how this might be done. If B_1 is the empirical science accepted at one time and B_2 is the empirical science accepted at a different time, then, according to this "quasi-intuitionist" interpretation, the very logical connectives would refer to provability in B_1, when used in B_1 and to provability in B_2 when used in B_2. The logical connectives would change meaning in a systematic way as empirical knowledge changed.

TRUTH

Suppose we formalize empirical science or some part of empirical science—that is, we formulate it in a formalized language L, with suitable logical rules and axioms, and with empirical postulates appropriate to the body of the theory we are formalizing. Following standard present-day logical practice, the predicate 'true' (as applied to sentences of L) would not itself be a predicate of L but would belong to a stronger metalanguage, ML. (Saul Kripke is exploring a method of avoiding this separation of object language and metalanguage, but this would not affect the present discussion.) This predicate might be defined (using the logical resources of ML but no descriptive vocabulary except that of L) by methods due to Tarski; or it might be taken as a primitive (undefined) notion of ML. In either case, we would wish all sentences of the famous form:

(T) 'Snow is white' is true if and only if snow is white

—all sentences asserting the equivalence of a sentence of L (pretend 'Snow is white' is a sentence of L) and the sentence of ML which says of that sentence that it is true—to be theorems of ML. (Tarski called this "Criterium W" in his *Wahrheitsbegriff*—and this somehow got translated into English as "Convention T." I shall refer to the requirement that all sentences of the form (T) be theorems of ML as Criterion T.)

What happens to 'true' if we reinterpret the logical connectives in the quasi-intuitionist manner just described? *It is possible to define 'true' exactly à la Tarski.* Only truth becomes "provability" (or, to be more precise, the double negation of provability—I will ignore this last subtlety). In short: the *formal* property of truth—the criterion of adequacy (Criterion *T*)—only *fixes* the extension of 'true' *if the logical connectives are classical.*

This means that we can extend the remark made in the section "Mathematical Intuitionism" (the first indented remark): even if the "natives" we are studying accept Criterion *T* in addition to accepting all classical tautologies, it does not follow just from that that their 'true' is the classical 'true'.

"Truth" (defined in the standard recursive way, following Tarski) becomes provability if the logical connectives are suitably reinterpreted. What does "reference" become?

On the Tarski definition of truth and reference,

(a) 'Electron' refers

is equivalent to

(b) There are electrons.

But if 'there are' is interpreted intuitionalistically, (b) asserts only

(c) There is a description D such that 'D is an electron' is provable in B_1.

And *this* could be true (for suitable B_1) even if there are no electrons! In short, the effect of reinterpreting the logical connectives intuitionalistically is that "existence" becomes intratheoretic. Actually, the effect is even more complicated than (c) if, in addition to understanding the connectives quasi-intuitionalistically (i.e., in the intuitionist manner, but with 'probability' relativized to B_1), we use the conjunction-negation translation to interpret the classical connectives, as suggested here. But this complication does not change the point just made: if the quantifiers, like the other logical connectives, are interpreted in terms of the notion of provability, then existence becomes intratheoretic.

CORRESPONDENCE THEORY OF TRUTH

Now, what I want to suggest (the reader has probably been wondering what all this is leading up to!) is that the effect of abandoning realism— that is, abandoning the belief in any describable world of unobservable things, and accepting, in its place, the belief that all the unobservable things (and, possibly, the observable things as well) spoken of in any generation's scientific theories, including our own, are mere theoretical conveniences, destined to be replaced and supplanted by quite different and unrelated theoretical constructions in the future—would not be a total scrapping of the predicates 'true' and 'refers' in their *formal* aspects. We could, as the above discussion indicates, keep formal semantics (including Tarski-type truth-definitions); even keep classical logic; and yet shift our notion of truth over to something approximating "warranted assertability." And I believe that this shift is what would in fact happen. (Of course, the formal details are only a rational reconstruction, and not the only possible one at that.)

Of course, there is no question of *proving* such a claim. It is speculation about human cognitive nature, couched in the form of a prediction about a hypothetical situation. But what makes it plausible is that just such a substitution—a substitution of 'truth within the theory' or 'warranted assertability' for the realist notion of truth—has always accompanied skepticism about the realist notion from Protagoras to Michael Dummett.

If this is right, then what is the answer to our original question: what is the relation between realist explanations of the scientific method, its success, its convergence, and the realist view of truth?

I remarked at the outset that realists claim to believe in something called a "correspondence theory of truth." But what is that?

If I am right, this is not a different *definition* of truth. There is only one way anyone knows how to define 'true' and that is Tarski's way. (Actually, as I mentioned earlier, Saul Kripke has a *new* way, but the difference from Tarski is inessential in this context, although it is important for the treatment of the antinomies.) But is Tarski's way "realist"?

Well, it depends. If the logical connectives are understood realistically ("classically," as people say), then a Tarski-type truth-definition is realist to at least this degree: satisfaction (of which truth is a special case) is a relation between words and things, more precisely, between formulas and finite sequences of things. ('Satisfies' is the technical term Tarski uses for what I have been calling "reference." For example, instead of saying " 'Electron' refers to electrons," he would say "The sequence of length one consisting of just x satisfies the formula 'electron (y)' if and only if x is an electron." 'Satisfies' has the technical advantage of applying to n-place formulas. For example, one can say that the sequence 'Abraham; Isaac' satisfies the formula 'x is the father of y'; but it is not customary to use 'refers' in connection with dyadic (or higher order) formulas, for example, to say that 'father of' refers to 'Abraham, Isaac'.) This certainly conforms to an essential part of the idea of a correspondence theory.

Still, one tends to feel dissatisfied with the Tarski theory as a reconstruction of the correspondence theory of truth even if the logical connectives are understood classically. I think that there are a number of sources of this dissatisfaction, which I have expressed in some of my writings, but it seems to me that Hartry Field put his finger on the main one: the fact that primitive reference (i.e., *satisfaction* in the case of primitive predicates of the language) is "explained" by a *list*.

But the list has a very special structure. Look at the following clauses from the definition of primitive reference:

1. 'Electron' refers to electrons.
2. 'Gene' refers to genes.
3. 'DNA molecule' refers to DNA molecules.

These are similar to the famous

4. 'Snow is white' is true if and only if snow is white

and the similarity is not coincidental: "true" is the 0-adic case of satisfaction (a formula is true if it has no free variables and the null sequence satisfies it). The criterion of adequacy (Criterion *T*) can be generalized as follows:

(Call the result 'Criterion *S*', '*S*' for satisfaction.) An adequate definition of 'satisfies-in-*L*' must yield as theorems all instances of the following schema: $\ulcorner p(x_1, \ldots, x_n) \urcorner$ is satisfied by the sequence $y_1, \ldots, y_n$ if and only if $P(y_1, \ldots, y_n)$.

Rewriting (1) above as

(1') 'Electron (x)' is satisfied by y_1 if and only if y_1 is an electron—which is how it would be written in the first place in Tarski-ese—we see that the structure of the list Field objects to is determined by Criterion *S*. But these criteria—*T*, or its natural generalization to formulas containing free variables, *S*—are determined by the formal properties we want the notions of truth and reference to have, by the fact that we *need*, for a variety of purposes, to have a predicate in our metalanguage that satisfies precisely the Criterion *S*. (This is why we would *keep* Criterion *S* even if we went over to an intuitionist or quasi-intuitionist meaning for the logical connectives.)

So I conclude that Field's objection fails, and that it is correct for the realist to define 'true' à la Tarski. Even though the notion of truth is derived, so to speak, by a "transcendental deduction,"[7] and Criterion *S* is justified similarly, satisfaction or reference is still, viewed from within our realist conceptual scheme, a relation between words and things—and one of explanatory value, as Boyd's argument shows.

Now that I have laid out this argument, let me give a shorter and sloppier argument to somewhat the same effect:

" 'Electron' refers to electrons"—*how else* should we say what 'electron' refers to from *within* a conceptual system in which 'electron' is a *primitive* term?

As soon as we analyze electrons—say "electrons are particles with such and such mass and negative unit charge"—we can say " 'electron' refers to particles of such and such mass and negative unit charge," but then "charge" (or whatever the primitive notions may be in our new theory) will be explained "trivially," that is, in accord with Criterion S. Given that Quinian predicament (Kantian predicament?) that there is a real world *but* we can only describe it in terms of our own conceptual system (well, we should use someone else's conceptual system?) is it surprising that *primitive* reference has this character of apparent triviality? I believe that Field would reply along the following lines: (1) None of this shows that truth and reference must be defined à la Tarski (i.e., defined à la Tarski for some preferred language and extended to other languages via translation); and (2) None of this shows that a "physicalistic" theory of reference (or at least of primitive reference, in some suitable sense) cannot be given. All we have shown is that a physicalistic theory of reference is not needed. But it might be (Field would argue) that one is possible, and that it might greatly enhance our understanding of the phenomenon of reference. After all, a physicalistic theory would not be incompatible with Tarskian truth/satisfaction-definitions.

Moreover, Field would argue, in accepting Boyd's account of realism, I have given Field himself strong ammunition. It has just been conceded that reference and truth are notions which enter into at least some causal explanations. In one sense, however, they are not causal explanatory notions; we still need them to do formal logic, for example, even if Boyd's causal explanations of the success of science are false. But if they enter into causal explanations at all, is it not possible that their causal-explanatory role justifies looking for a physicalistic account of what truth and reference *are*?[8]

NOTES

1. It will emerge that I think that realism is like an empirical hypothesis in that it could be false, and that facts are relevant to its support (or to criticizing it); but that doesn't mean that realism is scientific (in any standard sense of 'scientific'), or that realism is a hypothesis. I have discussed why belief in an external world and in other minds is not a "hypothesis." (See Hilary Putnam, *Mind, Language, and Reality*, Philosophical Papers, vol. 2 (Cambridge: Cambridge University Press, 1975), chaps. 1, 7.

2. R. Boyd, *Realism and Scientific Epistemology* (Cambridge: Cambridge University Press, forthcoming).

3. See Putnam, "What Theories Are Not," in my *Mathematics, Matter, and Method,* Philosophical Papers, vol. 1 (Cambridge: Cambridge University Press, 1975).

4. In "Language and Reality," in my *Mind, Language, and Reality.*

5. This was pointed out by Kurt Gödel in "On Intuitionistic Arithmetic and Number Theory," reprinted in *The Undecidable,* ed. Martin David (New York: Raven Press, 1965).

6. One problem with reformulating physics intuitionistically is that one cannot state laws such as Newton's law of gravity which assert that two different empirically given real numbers (the force of *A* on *B,* and the quotient of *g* times the product of the masses and the distance squared) are exactly equal in intuitionist mathematics. (See Putnam, "What is Mathematical Truth?" in *Mathematics, Matter and Method,* for an explanation.) But one *can* say that such a law holds to, say, thirty decimal places—and, if one does not expect the law to be retained in the long run anyway, and is not trying to "converge" by successive approximations to anything that is objectively "true," then one presumably would not mind weakening physical theory to this extent.

7. See H. Putnam, *Meaning and the Moral Sciences* (London: Routledge and Kegan Paul, 1978), lecture 1 (ed.).

8. For further discussion, see ibid., lecture 3 (ed.).

8

Experimentation and Scientific Realism

Ian Hacking

Experimental physics provides the strongest evidence for scientific realism. Entities that in principle cannot be observed are regularly manipulated to produce new phenomena and to investigate other aspects of nature. They are tools, instruments not for thinking but for doing.

The philosopher's standard "theoretical entity" is the electron. I will illustrate how electrons have become experimental entities, or experimenter's entities. In the early stages of our discovery of an entity, we may test hypotheses about it. Then it is merely a hypothetical entity. Much later, if we come to understand some of its causal powers and use it to build devices that achieve well-understood effects in other parts of nature, then it assumes quite a different status.

Discussions about scientific realism or antirealism usually talk about theories, explanation, and prediction. Debates at that level are necessarily inconclusive. Only at the level of experimental practice is scientific realism unavoidable—but this realism is not about theories and truth. The experimentalist need only be a realist about the entities used as tools.

A PLEA FOR EXPERIMENTS

No field in the philosophy of science is more systematically neglected than experiment. Our grade school teachers may have told us that scientific method is experimental method, but histories of science have become histories of theory. Experiments, the philosophers say, are of value only when they test theory. Experimental work, they imply, has no life of its own. So we lack even a terminology to describe the many varied roles of experiment. Nor has this one-sidedness done theory any good, for radically different types of theory are used to think about the same physical phenomenon (e.g., the magneto-optical effect). The philosophers of theory have not noticed this and so misreport even theoretical enquiry.

Different sciences at different times exhibit different relationships between "theory" and "experiment." One chief role of experiment is the creation of phenomena. Experimenters bring into being phenomena that do not naturally exist in a pure state. These phenomena are the touchstones of physics, the keys to nature, and the source of much modern technology. Many are what physicists after the 1870s began to call "effects": the photoelectric effect, the Compton effect, and so forth.[1] A recent high-energy extension of the creation of phenomena is the creation of "events," to use the jargon of the trade. Most of the phenomena, effects, and events created by the experimenter are like plutonium: they do not exist in nature except possibly on vanishingly rare occasions.[2]

In this paper I leave aside questions of methodology, history, taxonomy, and the purpose of experiment in natural science. I turn to the purely philosophical issue of scientific realism. Simply call it "realism" for short. There are two basic kinds: realism about entities and realism about theories. There is no agreement on the precise definition of either. Realism about theories says that we try to form true theories about the world, about the inner constitution of matter and about the outer reaches of space. This realism gets its bite from optimism: we think we can do well in this project and have already had partial success. Realism about entities—and I include processes, states, waves, currents, interactions, fields, black holes, and the like among entities—asserts the existence of at least some of the entities that are the stock in trade of physics.[3]

The two realisms may seem identical. If you believe a theory, do you not believe in the existence of the entities it speaks about? If you believe in some entities, must you not describe them in some theoretical way that you accept? This seeming identity is illusory. *The vast majority of experimental physicists are realists about entities but not about theories.* Some are, no doubt, realists about theories too, but that is less central to their concerns.

Experimenters are often realists about the entities that they investigate, but they do not have to be so. R. A. Millikan probably had few qualms about the reality of electrons when he set out to measure their charge. But he could have been skeptical about what he would find until he found it. He could even have remained skeptical. Perhaps there is a least unit of electric charge, but there is no particle or object with exactly that unit of charge. Experimenting on an entity does not commit you to believing that it exists. Only manipulating an entity, in order to experiment on something else, need do that.

Moreover, it is not even that you use electrons to experiment on something else that makes it impossible to doubt electrons. Understanding some causal properties of electrons, you guess how to build a very ingenious, complex device that enables you to line up the electrons the way you want, in order to see what will happen to something else. Once you have the right experimental idea, you know in advance roughly how to try to build the device, because you know that this is the way to get the electrons to behave in such and such a way. Electrons are no longer ways of organizing our thoughts or saving the phenomena that have been observed. They are now ways of creating phenomena in some other domain of nature. Electrons are tools.

There is an important experimental contrast between realism about entities and realism about theories. Suppose we say that the latter is belief that science aims at true theories. Few experimenters will deny that. Only philosophers doubt it. Aiming at the truth is, however, something about the indefinite future. Aiming a beam of electrons is using present electrons. Aiming a finely tuned laser at a particular atom in order to knock off a certain electron to produce an ion is aiming at present electrons. There is, in contrast, no present set of theories that one has to believe in. If realism about theories is a doctrine about the aims of science, it is a doctrine laden with certain kinds of values. If realism about entities is a matter of aiming electrons next week or aiming at other electrons the week after, it is a doctrine much more neutral between values. The way in which experimenters are scientific realists about entities is entirely different from ways in which they might be realists about theories.

This shows up when we turn from ideal theories to present ones. Various properties are confidently ascribed to electrons, but most of the confident properties are expressed in numerous different theories or models about which an experimenter can be rather agnostic. Even people in a team, who work on different parts of the same large experiment, may hold different and mutually incompatible accounts of electrons. That is because different parts of the experiment will make different uses of electrons. Models good for calculations on one aspect of electrons will be poor for others. Occasionally, a team actually has to select a member with a quite different theoretical perspective simply to get someone who can solve those experimental problems. You may choose someone with a foreign training, and whose talk is well-nigh incommensurable with yours, just to get people who can produce the effects you want.

But might there not be a common core of theory, the intersection of everybody in the group, which is the theory of the electron to which all the experimenters are realistically committed? I would say common lore, *not* common core. There are a lot of theories, models, approximations, pictures, formalisms, methods, and so forth involving electrons, but there is no reason to suppose that the intersection of these is a theory at all. Nor is there any reason to think that there is such a thing as "the most powerful nontrivial *theory* contained in the intersection of all the theories in which this or that member of a team has been trained to believe." Even if there are a lot of shared beliefs, there is no reason to suppose they form anything worth calling a theory. Naturally, teams tend to be formed from like-minded people at the same institute, so there is usually some real shared theoretical basis to their work. That is a sociological fact, not a foundation for scientific realism.

I recognize that many a scientific realism concerning theories is a doctrine not about the present but about what we might achieve, or possibly an ideal at which we aim. So to say that there is no present theory does not count against the optimistic aim. The point is that such scientific realism about theories has to adopt the Peircean principles of faith, hope, and charity. Scientific realism about entities needs no such virtues. It arises from what we can do at present. To understand this, we must look in some detail at what it is like to build a device that makes the electrons sit up and behave.

OUR DEBT TO HILARY PUTNAM

It was once the accepted wisdom that a word such as 'electron' gets its meaning from its place in a network of sentences that state theoretical laws. Hence arose the infamous problems of incommensurability and theory change. For if a theory is modified, how could a word such as 'electron' go on meaning the same? How could different theories about electrons be compared, since the very word 'electron' would differ in meaning from theory to theory?

Putnam saved us from such questions by inventing a referential model of meaning. He says that meaning is a vector, refreshingly like a dictionary entry. First comes the syntactic marker (part of speech); next the semantic marker (general category of thing signified by the word); then the stereotype (clichés about the natural kind, standard examples of its use, and present-day associations. The stereotype is subject to change as opinions about the kind are modified). Finally, there is the actual referent of the word, the very stuff, or thing, it denotes if it denotes anything. (Evidently dictionaries cannot include this in their entry, but pictorial dictionaries do their best by inserting illustrations whenever possible.)[4]

Putnam thought we can often guess at entities that we do not literally point to. Our initial guesses may be jejune or inept, and not every naming of an invisible thing or stuff pans out. But when it does, and we frame better and better ideas, then Putnam says that, although the stereotype changes, we refer to the same kind of thing or stuff all along. We and Dalton alike spoke about the same stuff when we spoke of (inorganic) acids. J. J. Thomson, H. A. Lorentz, Bohr, and Millikan were, with their different theories and observations, speculating about the same kind of thing, the electron.

There is plenty of unimportant vagueness about when an entity has been successfully "dubbed," as Putnam puts it. 'Electron' is the name suggested by G. Johnstone Stoney in 1891 as the name for a natural unit of electricity. He had drawn attention to this unit in 1874. The name was then applied to the subatomic particles of negative charge, which J. J. Thomson, in 1897, showed cathodes rays consist of. Was Johnstone Stoney referring to the electron? Putnam's account does not require an unequivocal answer. Standard physics books say that Thomson discovered the electron. For once I might back theory and say that Lorentz beat him to it. Thomson called his electrons 'corpuscles', the subatomic particles of electric charge. Evidently, the name does not matter much. Thomson's most notable achievement was to measure the mass of the electron. He did this by a rough (though quite good) guess at e, and by making an excellent determination of e/m, showing that m is about 1/1800 the mass of the hydrogen atom. Hence it is natural to say that Lorentz merely postulated the existence of a particle of negative charge, while Thomson, determining its mass, showed that there is some such real stuff beaming off a hot cathode.

The stereotype of the electron has regularly changed, and we have at least two largely incompatible stereotypes, the electron as cloud and the electron as particle. One fundamental enrichment of the idea came in the 1920s. Electrons, it was found, have angular momentum, or "spin." Experimental work by O. Stern and W. Gerlach first indicated this, and then S. Goudsmit and G. E. Uhlenbeck provided the theoretical understanding of it in 1925. Whatever we think, Johnstone Stoney, Lorentz, Bohr, Thomson, and Goudsmit were all finding out more about the same kind of thing, the electron.

We need not accept the fine points of Putnam's account of reference in order to thank him for giving us a new way to talk about meaning. Serious discussion of inferred entities need no longer lock us into pseudo-problems of incommensurability and theory change. Twenty-five years ago the experimenter who believed that electrons exist, without giving much credence to any set of laws about electrons, would have been dismissed as philosophically incoherent. Now we realize it was the philosophy that was wrong, not the experimenter. My own relationship to Putnam's account of meaning is like the experimenter's relationship to a theory. I do not literally believe Putnam, but I am happy to employ his account as an alternative to the unpalatable account in fashion some time ago.

Putnam's philosophy is always in flux. His account of reference was intended to bolster scientific realism. But now, at the time of this writing (July 1981), he rejects any "metaphysical realism" but allows "internal realism."[5] The internal realist acts, in practical affairs, as if the entities occurring in his working theories did in fact exist. However, the direction of Putnam's metaphysical antirealism is no longer scientific. It is not peculiarly about natural science. It is about chairs and livers too. He thinks that the world does not naturally break up into our classifications. He calls himself a transcendental idealist. I call him a transcendental nominalist. I use the word 'nominalist' in the old-fashioned way, not meaning opposition to "abstract entities" like sets, but meaning the doctrine that there is no nonmental classification in nature that exists over and above our own human system of naming.

There might be two kinds of internal realist, the instrumentalist about science and the scientific realist. The former is, in practical affairs where he uses his present scheme of concepts, a realist about livers and chairs but thinks that electrons are only mental constructs. The latter thinks that livers, chairs, and electrons are probably all in the same boat, that is, real at least within the present system of classification. I take Putnam to be an internal scientific realist rather than an internal instrumentalist. The fact that either doctrine is compatible with transcendental nominalism and internal realism shows that our question of scientific realism is almost entirely independent of Putnam's internal realism.

INTERFERING

Francis Bacon, the first and almost last philosopher of experiments, knew it well: the experimenter sets out "to twist the lion's tail." Experimentation is interference in the course of nature; "nature under constraint and vexed; that is to say, when by art and the hand of man she is forced out of her natural state, and squeezed and moulded."[6] The experimenter is convinced of the reality of entities, some of whose causal properties are sufficiently well understood that they can be used to interfere *elsewhere* in nature. One is impressed by entities that one can use to test conjectures about other, more hypothetical entities. In my example, one is sure of the electrons that are used to investigate weak neutral currents and neutral bosons. This should not be news, for why else are we (nonskeptics) sure of the reality of even macroscopic objects, but because of what we do with them, what we do to them, and what they do to us?

Interference and interaction are the stuff of reality. This is true, for example, at the borderline of observability. Too often philosophers imagine that microscopes carry conviction because they help us see better. But that is only part of the story. On the contrary, what counts is what we can do to a specimen under a microscope, and what we can see ourselves doing. We stain the specimen, slice it, inject it, irradiate it, fix it. We examine it using different kinds of microscopes that employ optical systems that rely on almost totally unrelated facts about light. Microscopes carry conviction because of the great array of interactions and interferences that are possible. When we see something that turns out to be unstable under such play, we call it an artifact and say it is not real.[7]

Likewise, as we move down in scale to the truly unseeable, it is our power to use unobservable entities that makes us believe they are there. Yet, I blush over these words 'see' and 'observe'. Philosophers and physicists often use these words in different ways. Philosophers tend to treat opacity to visible light as the touchstone of reality, so that anything that cannot be touched or seen with the naked eye is called a theoretical or inferred entity. Physicists, in contrast, cheerfully talk of observing the very entities that philosophers say are not observable. For example, the fermions are those fundamental constituents of matter such as electron neutrinos and deuterons and, perhaps, the notorious quarks. All are standard philosophers' "unobservable" entities. C. Y. Prescott, the initiator of the experiment described below, said in a recent lecture, that "of these fermions, only the t quark is yet unseen. The failure to observe $t\bar{t}$ states in e^+e^- annihilation at PETRA remains a puzzle."[8] Thus, the physicist distinguishes among the philosophers' "unobservable" entities, noting which have been observed and which not. Dudley Shapere has just published a valuable study of this fact.[9] In his example, neutrinos are used to see the interior of a star. He has ample quotations such as "neutrinos present the only way of directly observing" the very hot core of a star.

John Dewey would have said that fascination with seeing-with-the-naked-eye is part of the spectator theory of knowledge that has bedeviled philosophy from earliest times. But I do not think Plato or Locke or anyone before the nineteenth century was as obsessed with the sheer opacity of objects as we have been since. My own obsession with a technology that manipulates objects is, of course, a twentieth-century counterpart to positivism and phenomenology. Its proper rebuttal is not a restriction to a narrower domain of reality, namely, to what can be positivistically seen with the eye, but an extension to other modes by which people can extend their consciousness.

MAKING

Even if experimenters are realists about entities, it does not follow that they are right. Perhaps it is a matter of psychology: maybe the very skills that make for a great experimenter go with a certain cast of mind which objectifies whatever it thinks about. Yet this will not do. The experimenter cheerfully regards neutral bosons as merely hypothetical entities, while electrons are real. What is the difference?

There are an enormous number of ways in which to make instruments that rely on the causal properties of electrons in order to produce desired effects of unsurpassed precision. I shall illustrate this. The argument—it could be called the 'experimental argument for realism'—is not that we infer the reality of electrons from our success. We do not make the instruments and then infer the reality of the electrons, as when we test a hypothesis, and then believe it because it passed the test. That gets the time-order wrong. By now we design apparatus relying on a modest number of home truths about electrons, in order to produce some other phenomenon that we wish to investigate.

That may sound as if we believe in the electrons because we predict how our apparatus will behave. That too is misleading. We have a number of general ideas about how to prepare polarized electrons, say. We spend a lot of time building prototypes that do not work. We get rid of innumerable bugs. Often we have to give up and try another approach. Debugging is not a matter of theoretically explaining or predicting what is going wrong. It is partly a matter of getting rid of "noise" in the apparatus. "Noise" often means all the events that are not understood by any theory. The instrument must be able to isolate, physically, the properties of the entities that we wish to use, and damp down all the other effects that might get in our way. *We are completely convinced of the reality of electrons when we regularly set to build—and often enough succeed in building—new kinds of device that use various well understood causal properties of electrons to interfere in other more hypothetical parts of nature.*

It is not possible to grasp this without an example. Familiar historical examples have usually become encrusted by false theory-oriented philos-

ophy or history, so I will take something new. This is a polarizing electron gun whose acronym is PEGGY II. In 1978, it was used in a fundamental experiment that attracted attention even in *The New York Times*.
In the next section I describe the point of making PEGGY II. To do that, I
have to tell some new physics. You may omit reading this and read only
the engineering section that follows. Yet it must be of interest to know
the rather easy-to-understand significance of the main experimental
results, namely, that parity is not conserved in scattering of polarized
electrons from deuterium, and that, more generally, parity is violated in
weak neutral-current interactions.[10]

PARITY AND WEAK NEUTRAL CURRENTS

There are four fundamental forces in nature, not necessarily distinct.
Gravity and electromagnetism are familiar. Then there are the strong and
weak forces (the fulfillment of Newton's program, in the *Optics*, which
taught that all nature would be understood by the interaction of particles
with various forces that were effective in attraction or repulsion over
various different distances, i.e., with different rates of extinction).

Strong forces are 100 times stronger than electromagnetism but act
only over a miniscule distance, at most the diameter of a proton. Strong
forces act on "hadrons," which include protons, neutrons, and more
recent particles, but not electrons or any other members of the class of
particles called "leptons."

The weak forces are only 1/10,000 times as strong as electromagnetism, and act over a distance 100 times greater than strong forces. But
they act on both hadrons and leptons, including electrons. The most
familiar example of a weak force may be radioactivity.

The theory that motivates such speculation is quantum electrodynamics. It is incredibly successful, yielding many predictions better than
one part in a million, truly a miracle in experimental physics. It applies
over distances ranging from diameters of the earth to 1/100 the diameter
of the proton. This theory supposes that all the forces are "carried" by
some sort of particle: photons do the job in electromagnetism. We
hypothesize "gravitons" for gravity.

In the case of interactions involving weak forces, there are charged
currents. We postulate that particles called "bosons" carry these weak
forces.[11] For charged currents, the bosons may be either positive or negative. In the 1970s, there arose the possibility that there could be weak
"neutral" currents in which no charge is carried or exchanged. By sheer

analogy with the vindicated parts of quantum electrodynamics, neutral bosons were postulated as the carriers in weak neutral interactions.

The most famous discovery of recent high-energy physics is the failure of the conservation of parity. Contrary to the expectations of many physicists and philosophers, including Kant,[12] nature makes an absolute distinction between right-handedness and left-handedness. Apparently, this happens only in weak interactions.

What we mean by right- or left-handed in nature has an element of convention. I remarked that electrons have spin. Imagine your right hand wrapped around a spinning particle with the fingers pointing in the direction of spin. Then your thumb is said to point in the direction of the spin vector. If such particles are traveling in a beam, consider the relation between the spin vector and the beam. If all the particles have their spin vector in the same direction as the beam, they have right-handed (linear) polarization, while if the spin vector is opposite to the beam direction, they have left-handed (linear) polarization.

The original discovery of parity violation showed that one kind of product of a particle decay, a so-called muon neutrino, exists only in left-handed polarization and never in right-handed polarization.

Parity violations have been found for weak *charged* interactions. What about weak *neutral* currents? The remarkable Weinberg-Salam model for the four kinds of force was proposed independently by Stephen Weinberg in 1967 and A. Salam in 1968. It implies a minute violation of parity in weak neutral interactions. Given that the model is sheer speculation, its success has been amazing, even awe-inspiring. So it seemed worthwhile to try out the predicted failure of parity for weak neutral interactions. That would teach us more about those weak forces that act over so minute a distance.

The prediction is: slightly more left-handed polarized electrons hitting certain targets will scatter, than right-handed electrons. Slightly more! The difference in relative frequency of the two kinds of scattering is 1 part in 10,000, comparable to a difference in probability between 0.50005 and 0.49995. Suppose one used the standard equipment available at the Stanford Linear Accelerator Center in the early 1970s, generating 120 pulses per second, each pulse providing one electron event. Then you would have to run the entire SLAC beam for twenty-seven years in order to detect so small a difference in relative frequency. Considering that one uses the same beam for lots of experiments simultaneously, by letting different experiments use different pulses, and considering that no equipment remains stable for even a month, let alone twenty-seven years, such an experiment is impossible. You need enormously more electrons coming off in each pulse. We need between 1000 and 10,000 more electrons per pulse than was once possible. The first attempt used an instrument now called PEGGY I. It had, in essence, a high-class version of J. J. Thomson's hot cathode. Some lithium was heated and electrons were boiled off. PEGGY II uses quite different principles.

PEGGY II

The basic idea began when C. Y. Prescott noticed (by chance!) an article in an optics magazine about a crystalline substance called gallium arsenide. GaAs has a curious property; when it is struck by circularly polarized light of the right frequencies, it emits lots of linearly polarized electrons. There is a good, rough and ready quantum understanding of why this happens, and why half the emitted electrons will be polarized, three-fourths of these polarized in one directon and one-fourth polarized in the other.

PEGGY II uses this fact, plus the fact that GaAs emits lots of electrons owing to features of its crystal structure. Then comes some engineering—it takes work to liberate an electron from a surface. We know that painting a surface with the right stuff helps. In this case, a thin layer of cesium and oxygen is applied to the crystal. Moreover, the less air pressure around the crystal, the more electrons will escape for a given amount of work. So the bombardment takes place in a good vacuum at the temperature of liquid nitrogen.

We need the right source of light. A laser with bursts of red light (7100 Ångstroms) is trained on the crystal. The light first goes through an ordinary polarizer, a very old-fashioned prism of calcite, or Iceland spar[13]— this gives linearly polarized light. We want circularly polarized light to hit the crystal, so the polarized laser beam now goes through a cunning device called a Pockel's cell, which electrically turns linearly polarized photons into circularly polarized ones. Being electric, it acts as a very fast switch. The direction of circular polarization depends on the direction of current in the cell. Hence, the direction of polarization can be varied randomly. This is important, for we are trying to detect a minute asymmetry between right- and left-handed polarization. Randomizing helps us guard against any systematic "drift" in the equipment.[14] The randomization is generated by a radioactive decay device, and a computer records the direction of polarization for each pulse.

A circularly polarized pulse hits the GaAs crystal, resulting in a pulse of linearly polarized electrons. A beam of such pulses is maneuvered by magnets into the accelerator for the next bit of the experiment. It passes through a device that checks on a proportion of polarization along the way. The remainder of the experiment requires other devices and detectors of comparable ingenuity, but let us stop at PEGGY II.

BUGS

Short descriptions make it all sound too easy; therefore, let us pause to reflect on debugging. Many of the bugs are never understood. They are eliminated by trial and error. Let me illustrate three different kinds of bugs: (1) the essential technical limitations that, in the end, have to be factored into the analysis of error; (2) simpler mechanical defects you never think of until they are forced on you, and (3) hunches about what might go wrong.

Here are three examples of bugs:

1. Laser beams are not as constant as science fiction teaches, and there is always an irremediable amount of "jitter" in the beam over any stretch of time.

2. At a more humdrum level, the electrons from the GaAs crystal are back-scattered and go back along the same channel as the laser beam used to hit the crystal. Most of them are then deflected magnetically. But some get reflected from the laser apparatus and get back into the system. So you have to eliminate these new ambient electrons. This is done by crude mechanical means, making them focus just off the crystal and, thus, wander away.

3. Good experimenters guard against the absurd. Suppose that dust particles on an experimental surface lie down flat when a polarized pulse hits it, and then stand on their heads when hit by a pulse polarized in the opposite direction. Might that have a systematic effect, given that we are detecting a minute asymmetry? One of the team thought of this in the middle of the night and came down next morning frantically using anti-dust spray. They kept that up for a month, just in case.[15]

RESULTS

Some 10^{11} events were needed to obtain a result that could be recognized above systematic and statistical error. Although the idea of systematic error presents interesting conceptual problems, it seems to be unknown to philosophers. There were systematic uncertainties in the detection of right- and left-handed polarization, there was some jitter, and there were other problems about the parameters of the two kinds of beam. These errors were analyzed and linearly added to the statistical error. To a student of statistical inference, this is real seat-of-the-pants analysis with no rationale whatsoever. Be that as it may, thanks to PEGGY II the number of events was big enough to give a result that convinced the entire physics community.[16] Left-handed polarized electrons were scattered from deuterium slightly more frequently than right-handed electrons. This was the first convincing example of parity-violation in a weak neutral current interaction.

COMMENT

The making of PEGGY II was fairly nontheoretical. Nobody worked out in advance the polarizing properties of GaAs—that was found by a chance encounter with an unrelated experimental investigation. Although elementary quantum theory of crystals explains the polarization effect, it does not explain the properties of the actual crystal used. No one has got a real crystal to polarize more than 37 percent of the electrons, although in principle 50 percent should be polarized.

Likewise, although we have a general picture of why layers of cesium and oxygen will "produce negative electron affinity," that is, make it easier for electrons to escape, we have no quantitative understanding of why this increases efficiency to a score of 37 percent.

Nor was there any guarantee that the bits and pieces would fit together. To give an even more current illustration, future experimental work, briefly described later in this paper, makes us want even more electrons per pulse than PEGGY II can give. When the aforementioned parity experiment was reported in *The New York Times*, a group at Bell Laboratories read the newspaper and saw what was going on. They had been constructing a crystal lattice for totally unrelated purposes. It uses layers of GaAs and a related aluminum compound. The structure of this lattice leads one to expect that virtually all the electrons emitted would be polarized. As a consequence, we might be able to double the efficiency of PEGGY II. But, at present, that nice idea has problems. The new lattice should also be coated in work-reducing paint. The cesium-oxygen compound is applied at high temperature. Hence the aluminum tends to ooze into the neighboring layer of GaAs, and the pretty artificial lattice becomes a bit uneven, limiting its fine polarized-electron-emitting properties.[17] So perhaps this will never work. Prescott is simultaneously reviving a souped up new thermionic cathode to try to get more electrons. Theory would not have told us that PEGGY II would beat out thermionic

PEGGY I. Nor can it tell if some thermionic PEGGY III will beat out PEGGY II.

Note also that the Bell people did not need to know a lot of weak neutral current theory to send along their sample lattice. They just read *The New York Times*.

MORAL

Once upon a time, it made good sense to doubt that there were electrons. Even after Thomson had measured the mass of his corpuscles, and Millikan their charge, doubt could have made sense. We needed to be sure that Millikan was measuring the same entity as Thomson. Thus, more theoretical elaboration was needed, and the idea had to be fed into many other phenomena. Solid state physics, the atom, and superconductivity all had to play their part.

Once upon a time, the best reason for thinking that there are electrons might have been success in explanation. Lorentz explained the Faraday effect with his electron theory. But the ability to explain carries little warrant of truth. Even from the time of J. J. Thomson, it was the measurements that weighed in, more than the explanations. Explanations, however, did help. Some people might have had to believe in electrons because the postulation of their existence could explain a wide variety of phenomena. Luckily, we no longer have to pretend to infer from explanatory success (i.e., from what makes our minds feel good). Prescott and the team from the SLAC do not explain phenomena with electrons. They know how to use them. Nobody in his right mind thinks that electrons "really" are just little spinning orbs about which you could, with a small enough hand, wrap your fingers and find the direction of spin along your thumb. There is, instead, a family of causal properties in terms of which gifted experimenters describe and deploy electrons in order to investigate something else, for example, weak neutral currents and neutral bosons. We know an enormous amount about the behavior of electrons. It is equally important to know what does *not* matter to electrons. Thus, we know that bending a polarized electron beam in magnetic coils does not affect polarization in any significant way. We have hunches, too strong to ignore although too trivial to test independently: for example, dust might dance under changes of directions of polarization. Those hunches are based on a hard-won sense of the kinds of things electrons are. (It does not matter at all to this hunch whether electrons are clouds or waves or particles.)

WHEN HYPOTHETICAL ENTITIES BECOME REAL

Note the complete contrast between electrons and neutral bosons. Nobody can yet manipulate a bunch of neutral bosons, if there are any. Even weak neutral currents are only just emerging from the mists of hypothesis. By 1980, a sufficient range of convincing experiments had made them the object of investigation. When might they lose their hypothetical status and become commonplace reality like electrons?—when we use them to investigate something else.

I mentioned the desire to make a better electron gun than PEGGY II. Why? Because we now "know" that parity is violated in weak neutral interactions. Perhaps by an even more grotesque statistical analysis than that involved in the parity experiment, we can isolate just the weak interactions. For example, we have a lot of interactions, including electromagnetic ones, which we can censor in various ways. If we could also statistically pick out a class of weak interactions, as precisely those where parity is not conserved, then we would possibly be on the road to quite deep investigations of matter and antimatter. To do the statistics, however, one needs even more electrons per pulse than PEGGY II could hope to generate. If such a project were to succeed, we should then be beginning to use weak neutral currents as a manipulable tool for looking at something else. The next step toward a realism about such currents would have been made.

The message is general and could be extracted from almost any branch of physics. I mentioned earlier how Dudley Shapere has recently used "observation" of the sun's hot core to illustrate how physicists employ the concept of observation. They collect neutrinos from the sun in an enormous disused underground mine that has been filled with old cleaning fluid (i.e., carbon tetrachloride). We would know a lot about the inside of the sun if we knew how many solar neutrinos arrive on the earth. So these are captured in the cleaning fluid. A few neutrinos will form a new radioactive nucleus (the number that do this can be counted). Although, in this study, the extent of neutrino manipulation is much less than electron manipulation in the PEGGY II experiment, we are nevertheless plainly using neutrinos to investigate something else. Yet not many years ago, neutrinos were about as hypothetical as an entity could get. After 1946 it was realized that when mesons disintegrate giving off, among other things, highly energized electrons, one needed an extra nonionizing particle to conserve momentum and energy. At that time this postulated "neutrino" was thoroughly hypothetical, but now it is routinely used to examine other things.

CHANGING TIMES

Although realisms and antirealisms are part of the philosophy of science well back into Greek prehistory, our present versions mostly descend from debates at the end of the nineteenth century about atomism. Antirealism about atoms was partly a matter of physics; the energeticists thought energy was at the bottom of everything, not tiny bits of matter. It also was connected with the positivism of Comte, Mach, K. Pearson, and even J. S. Mill. Mill's young associate Alexander Bain states the point in a characteristic way, apt for 1870:

Some hypotheses consist of assumptions as to the minute structure and operation of bodies. From the nature of the case these assumptions can never be proved by direct means. Their merit is their suitability to express phenomena. They are Representative Fictions.[18]

"All assertions as to the ultimate structure of the particles of matter," continues Bain, "are and ever must be hypothetical. . . . The kinetic theory of heat serves an important intellectual function." But we cannot hold it to be a true description of the world. It is a representative fiction.

Bain was surely right a century ago, when assumptions about the minute structure of matter could not be proved. The only proof could be indirect, namely, that hypotheses seemed to provide some explanation and helped make good predictions. Such inferences, however, need never produce conviction in the philosopher inclined to instrumentalism or some other brand of idealism.

Indeed, the situation is quite similar to seventeenth-century epistemology. At that time, knowledge was thought of as correct representation. But then one could never get outside the representations to be sure that they corresponded to the world. Every test of a representation is just another representation. "Nothing is so much like an idea as an idea," said Bishop Berkeley. To attempt to argue to scientific realism at the level of theory, testing, explanation, predictive success, convergence of theories, and so forth is to be locked into a world of representations. No wonder that scientific antirealism is so permanently in the race. It is a variant on "the spectator theory of knowledge."

Scientists, as opposed to philosophers, did, in general, become realists about atoms by 1910. Despite the changing climate, some antirealist variety of instrumentalism or fictionalism remained a strong philosophical alternative in 1910 and in 1930. That is what the history of philosophy teaches us. The lesson is: think about practice, not theory. Antirealism about atoms was very sensible when Bain wrote a century ago. Antirealism about *any* submicroscopic entities was a sound doctrine in those days. Things are different now. The "direct" proof of electrons and the like is our ability to manipulate them using well-understood low-level causal properties. Of course, I do not claim that reality is constituted by human manipulability. Millikan's ability to determine the charge of the electron did something of great importance for the idea of electrons, more, I think, than the Lorentz theory of the electron. Determining the charge of something makes one believe in it far more than postulating it to explain something else. Millikan got the charge on the electron; but better still, Uhlenbeck and Goudsmit in 1925 assigned angular momentum to electrons, brilliantly solving a lot of problems. Electrons have spin, ever after. The clincher is when we can put a spin on the electrons, and thereby get them to scatter in slightly different proportions.

Surely, there are innumerable entities and processes that humans will never know about. Perhaps there are many in principle we can never know about, since reality is bigger than us. The best kinds of evidence for the reality of a postulated or inferred entity is that we can begin to measure it or otherwise understand its causal powers. The best evidence, in turn, that we have this kind of understanding is that we can set out, from scratch, to build machines that will work fairly reliably, taking advantage of this or that causal nexus. Hence, engineering, not theorizing, is the best proof of scientific realism about entities. My attack on scientific antirealism is analogous to Marx's onslaught on the idealism of his day. Both say that the point is not to understand the world but to change it. Perhaps there are some entities which in theory we can know about only through theory (black holes). Then our evidence is like that furnished by Lorentz. Perhaps there are entities which we shall only measure and never use. The experimental argument for realism does not say that only experimenter's objects exist.

I must now confess a certain skepticism, about, say, black holes. I suspect there might be another representation of the universe, equally consistent with phenomena, in which black holes are precluded. I inherit from Leibniz a certain distaste for occult powers. Recall how he inveighed against Newtonian gravity as occult. It took two centuries to show he was right. Newton's ether was also excellently occult—it taught us lots: Maxwell did his electromagnetic waves in ether, H. Hertz confirmed the ether by demonstrating the existence of radio waves. Albert A. Michelson figured out a way to interact with the ether. He thought his experiment confirmed G. G. Stoke's ether drag theory, but, in the end, it was one of many things that made ether give up the ghost. A skeptic such as myself has a slender induction: long-lived theoretical entities which do not end up being manipulated commonly turn out to have been wonderful mistakes.

NOTES

1. C. W. F. Everitt suggests that the first time the word 'effect' is used this way in English is in connection with the Peltier effect, in James Clerk Maxwell's 1873 *Electricity and Magnetism*, par. 249, p. 301. My interest in experiment was kindled by conversation with Everitt some years ago, and I have learned much in working with him on our joint (unpublished) paper, "Theory or Experiment, Which Comes First?"

2. Ian Hacking, "Spekulation, Berechnung und die Erschaffnung der Phänomenen," in *Versuchungen: Aufsätze zur Philosophie, Paul Feyerabends*, no. 2, ed. P. Duerr (Frankfort, 1981), 126-158.

3. Nancy Cartwright makes a similar distinction in her book, *How the Laws of Physics Lie* (Oxford: Oxford University Press, 1983). She approaches realism from the top, distinguishing theoretical laws (which do not state the facts) from phenomenological laws (which do). She believes in some "theoretical" entities and rejects much theory on the basis of a subtle analysis of modeling in physics. I proceed in the opposite direction, from experimental practice. Both approaches share an interest in real life physics as opposed to philosophical fantasy science. My own approach owes an enormous amount to Cartwright's parallel developments, which have often preceded my own. My use of the two kinds of realism is a case in point.

4. Hilary Putnam, "How Not to Talk About Meaning," "The Meaning of 'Meaning,'" and other papers in *Mind, Language and Reality*, Philosophical Papers, vol. 2 (Cambridge: Cambridge University Press, 1975).

5. These terms occur in, e.g., Hilary Putnam, *Meaning and the Moral Sciences* (London: Routledge and Kegan Paul, 1978), 123-130.

6. Francis Bacon, *The Great Instauration*, in *The Philosophical Works of Francis Bacon*, trans. Ellis and Spedding, ed. J. M. Robertson (London, 1905), 252.

7. Ian Hacking, "Do We See Through a Microscope?" *Pacific Philosophical Quarterly* 62 (1981): 305-322.

8. C. Y. Prescott, "Prospects for Polarized Electrons at High Energies," SLAC-PUB-2630, Stanford Linear Accelerator, October 1980, p. 5.

9. "The Concept of Observation in Science and Philosophy," *Philosophy of Science* 49 (1982): 485-526. See also K. S. Shrader-Frechette, "Quark Quantum Numbers and the Problem of Microphysical Observation," Synthese 50 (1982): 125-146, and ensuing discussion in that issue of the journal.

10. I thank Melissa Franklin, of the Stanford Linear Accelerator, for introducing me to PEGGY II and telling me how it works. She also arranged discussion with members of the PEGGY II group, some of whom are mentioned below. The report of experiment E-122 described here is "Parity Non-conservation in Inelastic Electron Scattering," C. Y. Prescott et al., in *Physics Letters*. I have relied heavily on the in-house journal, the *SLAC Beam Line*, report no. 8, October 1978, "Parity Violation in Polarized Electron Scattering." This was prepared by the in-house science writer Bill Kirk.

11. The odd-sounding bosons are named after the Indian physicist S. N. Bose (1894-1974), also remembered in the name "Bose-Einstein statistics" (which bosons satisfy).

12. But excluding Leibniz, who "knew" there had to be some real, natural difference between right- and left-handedness.

13. Iceland spar is an elegant example of how experimental phenomena persist even while theories about them undergo revolutions. Mariners brought calcite from Iceland to Scandinavia. Erasmus Bartholinus experimented with it and wrote it up in 1609. When you look through these beautiful crystals you see double, thanks to the so-called ordinary and extraordinary rays. Calcite is a natural polarizer. It was our entry to polarized light which for three hundred years was the chief route to improved theoretical and experimental understanding of light and then electromagnetism. The use of calcite in PEGGY II is a happy reminder of a great tradition.

14. It also turns GaAs, a 3/4 to 1/4 left-hand/right-hand polarizer, into a 50-50 polarizer.

15. I owe these examples to conversation with Roger Miller of SLAC.

16. The concept of a "convincing experiment" is fundamental. Peter Gallison has done important work on this idea, studying European and American experiments on weak neutral currents conducted during the 1970s.

17. I owe this information to Charles Sinclair of SLAC.

18. Alexander Bain, *Logic, Deductive and Inductive* (London and New York, 1870), 362.

9

Explanation and Realism

Clark Glymour

I

One way to argue to a theory is to show that it provides a good explanation of a body of phenomena and, indeed, that it provides a better explanation than does any available alternative theory. This pattern of argument is not bounded by time or by subject matter. One can find such arguments in sociology, in psychometrics, in chemistry, and in astronomy, in the time of Copernicus and in the most recent scientific journals. The goodness of explanations is a ubiquitous criterion; in every scientific subject it forms one of the principal standards by which we decide what to believe. The ambition of philosophy of science is, or ought to be, to obtain from the literature of the sciences a plausible and precise theory of scientific reasoning and argument: a theory that will abstract general patterns from the concreta of debates over genes and spectra and fields and delinquency. A philosophical understanding of science should, therefore, give us an account of what explanations are and of why they are valued but, most important, it should also provide us with clear and plausible criteria for comparing the goodness of explanations. One half of the subject of this essay concerns a fragment of such a theory. I will try to describe, without gratuitous formality, some features that generate clear and powerful criteria for judging the goodness of explanations. That is half of my subject; the remainder concerns what such criteria determine about what we ought to believe. Both halves are prompted by Bas van Fraassen's recent and delightful book, *The Scientific Image*.

Wherever theorists have postulated features of the world that could not, at the time, be observed, debates have erupted over the scientific credentials of beliefs in the unobserved or unobservable. The scientific debates have more often than not been best articulated in throwaway lines: "If I were the Master," wrote Dumas in the 1830s,[1] "I would ban the word 'atom' from chemistry, for it goes beyond all experience, and never, in chemistry, should we go beyond experience." And in our own time, we have B. F. Skinner's argument by rhetorical question: When one has a secure equational linkage between observed variables, why introduce gratuitous unobserved variables? We also have more considered, philosophical discussions that lead to the same opinion, namely, that there is no warrant to be found in scientific observations for conclusions that concern unobserved or unobservable features of the world.

Call those who hold such views "antirealists," and those who hold the contrary view, "realists." Johannes Kepler, J. Dalton, and C. Spearman were realists. Andreas Osiander, F. Dumas, and Skinner were (and in the last case remains) antirealists. The antirealist case has been made with care and with something approaching logical precision by several writers, but still nowhere better than by C. G. Hempel who argued as follows:[2] Make the idealization which supposes that all the evidence pertinent to our theories can be formulated using only a part of the terminology of science. Call this part the "observational vocabulary." Then full-fledged theories whose vocabulary goes beyond this observational fragment seem to have no epistemological advantage over the collection of their consequences that can be stated in observational terms alone. For example, Hempel views explanation as a kind of deductive systematization, and the "observational consequences" of a theory are as well systematized—by Hempel's criteria anyway—as is the theory itself. (It must be noted that Hempel himself did not draw the antirealist conclusion, for he hoped that theories might be shown to afford an "inductive systematization" unobtainable without them.) Further, both Duhem and Quine[3] have argued that once we entertain hypotheses that transcend all possible experience, we enter into the realm of convention and arbitrariness, for the evidence and the canons of scientific method can only determine what we ought to believe about the observable, and the unobservable is indeterminate: a plethora of alternative conjectures will explain the evidence equally well. Claims of the same sort were made against Dalton's theory by his anti-atomist opponents—the properties of atoms, they said, are indeterminate and arbitrary.

Duhem, Quine, and others are right enough if we allow science only a sufficiently impoverished collection of principles of assessment and inference. But even the fragmentary criteria for comparing scientific explanations which I am able to display are strong enough to defeat their anti-realist arguments. For although the criteria make no use of the notion of observability, they yield the result that sometimes the best explanation *does* go well beyond what is observed or observable. Or, to put it in something more like Hempel's fashion, the criteria yield the result that sometimes a theory with "nonobservation terms" has explanatory virtues that are unobtainable without such terms. Moreover, the criteria for comparing explanations do not, in leading us beyond the observable, enmesh us in interdeterminacy. On the contrary, in some contexts they suffice to determine a unique, best theory which explains the phenomena. Further, there is nothing about these criteria for the goodness of explanations which requires or presupposes that they be applied only to theories that contain claims about what is unobservable. The very same criteria are used to determine the best of competing theories about observable features of the world. The same criteria sometimes determine—at least as far as I understand the notion of 'observability'—that theories that are confined to observable features of the world are better explainers than competitors that postulate unobservable features. This all seems to me exactly as it should be. The result is simply that the same features of inference which lead to general conclusions about the observable also lead in other contexts to determinate conclusions about the unobservable. In consequence, I believe there are only two ways to maintain the antirealist position: either by impoverishing (perhaps I should say emasculating) the methods of science, and disallowing altogether explanatory criteria such as those I will describe, or by arbitrarily (and vaguely) restricting the scope of application of such principles to the realm of the observable.

II

In discussions of scientific explanation, and of the grounds such explanations may afford for belief in unobserved features of the world, it is important to keep actual cases in mind. Eventually, I will try to show that all of the following cases have something important in common:

1. The Copernican explanation of the regularities of the superior planets, in particular the regularity noted by Ptolemy, that if in a whole number of solar years a superior planet goes through a whole number of oppositions and also a whole number of revolutions of longitude, then the number of solar years equals the number of oppositions plus the number of revolutions of longitude. Copernican theory explains the regularity by noting that the number of solar years is equal to the number of revolutions the earth has made around the sun, and that the number of

oppositions is equal to the number of times the faster moving earth has overtaken the superior planet moving more slowly in an orbit outside the earth's orbit, and also that the number of revolutions of longitude equals the number of revolutions the superior planet has made about the sun.

2. The Daltonian explanation of the law of definite proportions. The law asserts that in any two cases in which quantities of the same pure chemical reagents combine to produce quantities of the same products, the ratios of the combining weights of the reactants are the same. Dalton's explanation is that any sample of a pure chemical substance consists of molecules, each having the same weight as any other, and that each molecule of a given kind is composed of the same numbers of atoms of various kinds. All atoms of the same kind have the same weight, and a chemical reaction is just a process in which the constituent atoms of the molecules in the reagents are recombined to form molecules of the products. Thus, the ratios of the numbers of molecules of the various reactants that combine are invariant and characteristic of the reaction. Two different samples of reactants with different weights will differ in the number of molecules they contain, but since the weight of any pure sample is just the sum of the weights of its constituent molecules, the ratio of the weights of the reactants will equal the ratio of the number of molecules they severally contain.

3. The general relativistic explanation of the anomalous motion of Mercury's perihelion, and of the deflection of starlight passing near the limb of the sun. The explanations go roughly as follows: for weak gravitational fields, such as those of the planets at astronomical distances, the theory of relativity closely approximates Newtonian gravitational theory. Hence, in the case of Mercury as well as in the case of the starlight, the only significant contribution to the gravitational field is that of the sun, which is known of to be close to spherical. Therefore, the gravitational field is, to good approximation, the gravitational field of a single mass point. From the field equations of the theory, the gravitational field, in empty space, of a single mass point is uniquely determined and describes the geometry of the surrounding space-time. The equations of motion of the theory imply that light rays move on null geodesics of this geometry and massive particles move on timelike geodesics. Light from a star may be described as a ray, and the planet Mercury reacts to gravity (to good approximation) as a point mass. Given the initial conditions, the phenomena follow.

4. Spearman's explanation of the correlations among intelligence tests. Although the history is complicated, it is probably fair to say that Spearman invented factor analysis when, in 1904, he published two papers, one on the measurement of correlations, and another on the measurement of intelligence.[4] Spearman obtained several assessments of popula-

tions of school children, including assessments of "intelligence" and of sensory discrimination. Analyzing his data from 1904 and later, he argued in his *Analysis of Human Abilities*, that for any four such measures, say X_1, X_2, X_3, X_4, the correlations among the measures have vanishing "tetrad differences." That is to say, empirically it is found that for any such quadruple of measures for his samples:

$$\rho_{12}\rho_{34} = \rho_{13}\rho_{24}$$
$$\rho_{13}\rho_{24} = \rho_{14}\rho_{23}$$

where ρ_{ij} is the correlation between X_i and X_j. Spearman claimed that the best explanation of these equations is that there is a common factor which determines the outcomes of scores on all of the measures, and he labeled that factor "general intelligence."

These aforementioned cases are not essential ones for any theory of explanation, but as cases they have important virtues. They represent a range of subject matters, they are explanations that have a genuine historical role, and they were not used to satisfy idle curiosity—as with questions about soap bubbles—but were instead crucial pieces of the arguments given for the respective theories. They also have no uniform connection with distinctions between what is observable and what is not. The Copernican theory postulated features—the motions of the planets in three-dimensional space—that could not be observed in the sixteenth century. So did competing theories. Yet, some of these features might arguably be regarded as observable today. Dalton's atomic theory certainly did concern features of the world that have served almost as paradigm cases of the unobservable. I have no idea whether the metric field of space-time ought to count as observable or as unobservable. Spearman's "general intelligence" is surely an unobservable feature of persons—more carefully, since one may doubt there is any such thing, if there were a factor such as Spearman's general intelligence, it would be unobservable. The phenomena explained in these cases can be viewed sometimes as regularities, sometimes as particular events, or sequences of events, and sometimes as both. The regularity of the superior planets is indeed a regularity, but Copernician theory also explains all of the instances of the phenomenon. The same holds of definite proportions and the motion of Mercury.

The explanation of the deflection of light was, in fact, the explanation of a few events, but there is a corresponding regularity that the theory explains. Spearman's equations neither describe particular events nor express generalizations; I suppose they are best understood as expressing particular features of the population tested. Finally, for each of these explanations, there was available an alternative theory generating competing explanations. The gravamen of the scientific arguments was that the explanations cited were *better* than any others available. Copernican explanations were compared with Ptolemaic explanations, general relativistic explanations with classical ones, and one-factor accounts of correlations were compared with multifactor accounts.

III

I view one of the chief goals of a philosophical account of aspects of scientific explanation as that of providing a canon or canons for the assessment of scientific theories. To the best of my knowledge, the philosophical literature on scientific explanation provides no clear criteria that can serve to explain why the four cases I have cited might have been viewed as good explanations or, in particular, why they should have been viewed as better than competing explanations. Nonetheless, what I shall have to offer by way of criteria is closely connected with some philosophical articulations of the notion of scientific explanation.

Philosophical theories of scientific explanation can be roughly divided into three types: purely logical theories, which analyze explanation solely in terms of logical relations and truth conditions; theories with extra objective structure, which impose objective conditions on explanation beyond those of truth and logical structure; and theories with extra subjective structure, which impose on explanations psychological conditions of belief, interest, and so forth.[5]

The apparent aim of Hempel and Oppenheim[6] was to provide an account of the logical structure of 'explains' in much the way that the logical tradition of Frege, Russell, Whitehead, and Hilbert had provided accounts of the logical structure of "is a proof of." The conditions given by Hempel and by Oppenheim were not intended as an analysis of when it is appropriate to say that someone has explained something to someone; instead, they were intended to specify the logical structure that fully explicit, nonstatistical explanations in the natural sciences would typically have if there were any, and which actual explanations in the natural sciences typically abbreviate. If Hempel and Oppenheim had been successful in their aim, their criteria would have formulated a critical standard for judging theories and hypotheses. On the basis of logical structure alone, it would be fully determinate what singular sentences a given theory could potentially explain, and what sentences it could not possibly explain. Combined with a further account of how to use information about what theories can and cannot explain in order to assess those theories, a logical account such as Hempel and Oppenheim's promised to provide us with an understanding of how explanation is used in deciding what to accept and believe.

Hempel and Oppenheim's account of the logical structure of deductive explanations turned out to be altogether unequal to its task. In effect, given an *arbitrary* true (but not logically true) sentence E and an *arbitrary* universally quantified sentence S which is not a logical truth, there exists a logical consequence of S and a true singular sentence C, such that the logical consequence of S and C together constitute the premises of a potential explanation of E.[7] In effect, anything explains anything. What was important in Hempel and Oppenheim's work was not the execution but the vision. If one shared the vision, the natural response to the failure of their analysis is to attempt another of the same kind, either in entirely logical terms or, perhaps, with extra structure.

Several purely logical replacements for Hempel and Oppenheim's analysis have been published. While not all of them are trivial, they are, without exception, far too weak to provide useful criteria for theory assessment. Perhaps the best attempt is David Kaplan's,[8] who uses conditions of truth as well as logical structure to constrain explanation.

Accounts such as Kaplan's are deficient in tolerating as explanations a range of cases that we would dismiss as not explanatory. For any explanation of the form:

$$\frac{\forall x\, Dx}{Da} \tag{1}$$

if P is any property which the individual a also has, then

$$\frac{\forall x\, (Px \to Dx)}{Pa} \tag{2}$$
$$\frac{Pa}{Da}$$

is an explanation meeting Kaplan's criteria and likewise various generalizations of Kaplan's criteria. Again, for any explanation of the form:

$$\frac{\forall x\, (Fx \to Gx)}{Fa} \tag{3}$$
$$\frac{Fa}{Ga}$$

if P is any property which the individual a also has, then

$$\frac{\forall x\, (Fx\ \&\ Px \to Gx)}{Fa\ \&\ Pa} \tag{4}$$
$$\frac{Fa\ \&\ Pa}{Ga}$$

is likewise an explanation meeting Kaplan's and related criteria. Thus, to use an example of Henry Kyburg's, we can explain the fact that a sample of table salt dissolves in water by citing the (true) claim that the sample of table salt has been hexed, and the (true) generalization that all hexed table salt dissolves in water.

One sensible response to these difficulties is that they show nothing more than the incompleteness of the analysis of explanation. Thus, to an account of Kaplan's kind, restricting the logical form of explanations, one should add criteria for comparing explanations with the appropriate logical form. For example: when an explanation of form (1) exists, any explanation of form (2) is defeated. Alternatively, one might reconstruct the logical conditions for explanation so that (2) and (4) will not count as explanations at all. Wesley Salmon proposes to regard deductive explanation as a limiting case of statistical explanation:[9] explaining a single event consists in assigning it to one member of a statistically homogeneous partition of the largest class of events that can be so partitioned. Very cleverly, it follows that when (1) obtains, (2) does not meet the conditions for explanation (unless P and D are coextensive), and analogously with (3) and (4).

None of this work on explanation provides any criteria that can help to account for the cases mentioned earlier, or for others like them. Ptolemaic and Copernican astronomical theories each provided deductions of sentences about the positions of the planets from general lawlike sentences; if that is what is required for potential explanation, then both theories provided it. The atomic theory explained each instance of the law of definite proportions, but so far as these logical criteria are concerned, equally good explanations were obtained from the view, popular in the nineteenth century, that one ought to abjure atoms and simply make tables of combining weights. Similar remarks hold for the other cases. Salmon's account of explanation, it is fair to say, is such that we simply do not know how to apply it in these sorts of contexts.

A simple way to see the limitations of the deductive criteria described so far is to consider what it would be to explain a regularity rather than a single event. One naturally supposes that to explain a regularity is to explain every instance of it, but if no more is required, then every regularity explains itself. If we require further that what explains a regularity not be logically implied by the regularity itself, we are still no better off, as Hempel and Oppenheim noted, since it will remain the case that by these criteria any regularity can be explained by conjoining that regularity with any other regularity logically independent of it.

I do not know of any writer who has attempted to account for the explanation of regularities in logical terms alone. Inevitably, considerations other than truth and logical form enter into the criteria, and that is quite proper and harmless so long as the further structure does not defeat the very goal of an account of explanation—namely, to help us to understand how explanation can be and is used to assess our hypotheses. Three recent discussions of explanation that appeal to extra structure are especially valuable, not so much for any technical advance they make, but for pointing to features of explanation easily overlooked, and for at least suggesting that such features might be formally tractable. One is Michael Friedman's proposal that what good theoretical explanations do is to somehow reduce the number of hypotheses that must be accepted independently of one another.[10] Another is Robert Causey's careful analysis of the structure of intertheoretical reductions,[11] with its emphasis on the identity of objects and properties described and postulated by the hypotheses to be explained with complexes of objects and properties postulated by the theory that provides the explanation. A third is Baruch Brody's attempt to bring Aristotle back to scientific explanation,[12] and to take seriously the idea that good scientific explanations show that the event or regularity to be explained is *necessary*, not just a necessary consequence of theoretical assumptions. None of these three accounts suffices to treat the sorts of cases I described earlier, and only Friedman's really constitutes an attempt to do so. However, Friedman's technical account came apart; Causey's analysis has technical difficulties and does not provide applicable comparative criteria; and Brody's account of explanation can be applied to compare competing claims to explain a phenomenon only in those situations where we know, independently, something about what properties are essential and what properties are accidental. Even so, in each of these three cases the motives seem sound in their own way, and the scheme that I present below can be viewed as an attempt to unify aspects of each of them.

IV

One of the things that many explanations seem to do, and which appeals to certain intuitions about explanation, is to demonstrate that one phenomenon is really just a variant of, a different manifestation of, another phenomenon. In some sciences, the explanations provided in textbooks are very largely of this kind. In classical thermodynamics, for example, the standard exercise repeated throughout textbooks of physical chemistry consists in using thermodynamic principles to derive empirical regularities from one another. The first and second laws are used to derive Joule's law from the ideal gas law, or to derive from the gas law the fact that specific heat at constant pressure minus specific heat at constant volume is proportional to the gas constant. Intuitively, an explanation produces a form of understanding if it shows us that the phenomenon we want explained is a manifestation of a different phenomenon we already know about. The understanding consists in a grasp of the unity of pattern or substance behind the apparently disparate phenomena. In physical contexts, the mechanics of such explanations is familiar to anyone who has suffered through problem sets. In the thermodynamic case again, one starts, say with $C_p - C_v = R$ and explains this relation by relating a subformula of it (e.g., $C_p - C_v$) to thermodynamic quantities, applying the thermodynamic laws to these quantities, and deriving the result that $C_p - C_v = PV/nT$.[13]

As a first approximation to a formal account of this feature of explanation, consider theories represented in the predicate calculus. For convenience, let us suppose that all sentences considered are in prenex form. Let a *subformula* of a sentence S be any well-formed subformula occurring in the matrix of the prenex form which is not logically equivalent to the entire sentence. A subformula of a sentence may have several occurrences in a sentence. By a term in S we will mean any term occurring in the matrix of S. Let us say that where T, H, and K are sentences, T *explains H as a result of K* if for (nonvalid) subformulas $H_1, \ldots, H_n$ of H, no one of which occurs in H as a subformula of any of the others, and terms $t_1, \ldots, t_k$ of H, not occurring in $H_1, \ldots, H_n$, there are formulas $K_1, \ldots, K_n$, terms $s_1, \ldots, s_k$, and subtheories $T_1, \ldots, T_{n+k}$, of T such that

(i) $T_i \vdash H_i \leftrightarrow K_i$
(ii) $T_{n+j} \vdash t_j = s_j$
(iii) $T_i \vdash H$ for any $i < n+k$
(iv) If $H(H_i t_j / K_i s_j)$ is the result of simultaneously substituting, for all i and j, K_i for every occurrence of H_i in H, and s_j for every occurrence of t_j in H, then $K \vdash H(H_i t_j / K_i s_j)$.

The third condition above is both natural and necessary to eliminate certain spurious explanations. For example, one could not use any theory including the ideal gas law to explain that very law as a form of the relation $T = T$ (where T is the temperature) by starting with $PV = nRT$, and noting that from the theory it follows that $PV/nR = T$ and that on substituting T for PV/nR, in the ideal gas law, one obtains $T = T$. Such an explanation is excluded by condition (iii). The condition that a term substituted for not occur as part of a subformula substituted for is perhaps unduly restrictive, but is intended to avoid complications.

In practice, we do not usually consider first-order formulas but rather equations representing theories which are to provide the explanations. In such a context, T explains H as a result of K, where $H(X_1, \ldots, X_n) = 0$ and $K(Y_1, \ldots, Y_m) = 0$ are equations and T a system of equations, if there are consequences T_i of T, not entailing H, and each T_i entails an equation between a combination of the Xs occurring in H and a combination of Ys, and when the combinations of Ys are substituted in H for the corresponding combination of Xs, one obtains an equation that is satisfied whenever K is.

As I have presented these conditions, they do not satisfy the condition of logical equivalence; that is to say, T can explain H as a result of K while for logically equivalent formulas T', H', and K', T' fails to explain H' as a result of K'. Of course, what I mean is that T explains H as a result of K if each of these sentences has equivalents meeting the formal conditions. In fact, the equivalence is generally much wider than logical equivalence, for we count an equation H as explained as a result of K if we can find mathematical equivalents of H and K meeting the conditions. Doubtless, the formal proposal has other defects as well; it is, at best, a first try. I hope that veterans of the textbook tradition in physics and chemistry at least will be tingled by a reminiscence that much of the work in the explanations presented and the problems assigned was to find the right representations of equations, and the right subtheories, in order to derive one thing from another.

A theory may entail each member of a class of regularities without explaining any one of them as a result of any other of them. If, for example, the regularities in question are logically (mathematically) independent, the theory consisting of their conjunction will have this property. Thus, to borrow an example of Friedman's, the theory consisting of the conjunction of Kepler's laws, Boyle's law, Galileo's law of falling bodies, and the law of the pendulum will not explain any one of these regularities as the result of any other. Again, the conjunction of the additivity of masses and the law of definite proportions does not explain either of these regularities by the other. By contrast, the atomic theory explains the law of definite proportions as a result of the additivity of masses, and that unification is one of the great virtues of the theory. In some contexts in which theories are being assessed and compared, there is a reasonably well-established collection of logically independent empirical regularities pertinent to the assessment. Contending theories, such as the atomic theory and the theory of equivalents, may well entail a common class of regularities, but the classes of regularities pertinent to the respective theories need not be coextensive: the additivity of masses has nothing to do with the theory of equivalents but a great deal to do with the theory of atoms. In general, we prefer theories that explain entailed regularities as the result of other established regularities to theories that do not. More exactly, I propose the following rather modest principle of comparison:

1. *Ceteris Paribus*, if T and Q are theories and for every established pair of regularities, H, K, such that Q explains H as a result of K, T also explains H as a result of K, but there exist established regularities, L, J, such that T explains J as a result of L but Q does not explain J as the result of any other established regularity, T is preferable to Q.

The statement is cumbersome, but the idea is almost trivial, and it seems to me to be part of what is correct about Friedman's suggestion that good theories reduce the number of independently acceptable hypotheses. It might be interesting to explore further principles of comparison that would extend to circumstances in which competing theories explain a common set of regularities as the result of other regularities, but disagree about which regularities are the result of which other regularities. I leave the matter aside here, except for a special case to be discussed later.

There are two points to emphasize. First, according to principle 1, certain theories entailing empirical regularities may be entirely preferable to the conjunction of those regularities alone, and may be preferable as well to the conjunction of the regularities entailed with any regularities from which these are claimed to result. The atomic theory illustrates the point. Second, according to principle 1, Copernican theory is preferable to its principal rival, Ptolemaic astronomy, and Spearman's explanation of the correlation equations among four measured variables in terms of a single common factor is preferable to many alternative explanations that equally save the phenomena. To see this, it will help to return to Brody's idea.

At least part of Brody's idea is that we may explain a state of affairs by showing it to be necessary, and that such a demonstration is to be given by showing that the state of affairs is a logical consequence of essential properties of the entities concerned, and of laws governing these essential properties. I suggest a different kind of demonstration of necessity: we may show that a regularity has a kind of necessity by explaining it as a result of a necessary truth. In particular, in the schema for 'T explains H as a result of K', nothing prohibits that K be a logical or mathematical, and thus a necessary, truth. Logical and mathematical truths, in general, may be considered regularities, and they may certainly be established and used in the assessment of theories. In the case of Copernican positional astronomy, and in the case of Spearman's tetrad equations, regularities or equations of an empirical kind are explained as the result of mathematical truths.

It is a mathematical truth that if two objects move in closed orbits about a common center, one orbit inside the other, and the inner object moves faster than the outer object, then the number of revolutions of the inner object equals the number of revolutions of the outer object plus the number of times the inner object has overtaken the outer object, whenever all of these numbers are whole. According to Copernican theory, the Earth moves in an orbit interior to the orbit of any superior planet, and moves faster in its orbit than does a superior planet. The number of solar years equals the number of revolutions of the Earth in its orbit; the number of oppositions or cycles of anomaly equals the number of times the superior planet is overtaken by the Earth, and (over any long period) the number of revolutions of longitude of a superior planet equals the number of revolutions in its orbit. Thus, the planetary regularity noted by Ptolemy is explained by Copernican theory as the result of a mathematical truth. It is not explained in Ptolemaic theory as the result of any other independently established regularity.

Spearman assumes (tacitly) that the relations between measured variables and unmeasured factors are linear. Assuming that measurement error averages to zero and that errors in measured variables are uncorrelated with one another, it follows that for standardized variables, the correlation between two variables is equal to the product of the linear coefficients relating each variable of the pair to the common factor. Thus, with Spearman's causal scheme

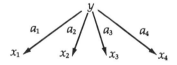

we have that

$$\rho_{ij} = a_i a_j.$$

On substituting the right-hand side of these equations into Spearman's supposed empirical equations:

$$\rho_{12}\rho_{34} = \rho_{13}\rho_{24}$$
$$\rho_{13}\rho_{24} = \rho_{14}\rho_{23}$$

we obtain instances of trivial truths of real algebra. Thus, Spearman's tetrad equations are explained as results of mathematical truths. Typically, other arrangements of unmeasured factors can be found which are consistent with the tetrad equations but do not explain the tetrad equations as the results of mathematical truths.

One might wonder whether, in the Copernican and psychometrical cases, an empirical regularity has been entirely explained as the result of necessary truths, for the Copernican relations between numbers of orbits and revolutions in longitude and solar years, and so forth are at least arguably not necessary truths. Again, psychometric equations relating correlation coefficients to "factor leadings" scarcely seem to be necessary truths. There are cases and contexts, however, in which an empirical relation seems literally to be explained *as* a necessary truth. I have in mind cases in which a regularity is explained as the result of a mathematical truth, and the equations used to connect quantities in the regularity with other quantities (as in [ii] and [iii] of the scheme) are identifications, as Professor Causey requires for microreductions. One case in which there is a tradition—correct or not—providing such an interpretation is that of general relativity. Both the motion of Mercury and the motion of light rays near the limb of the sun provide special cases of the general relativistic law of motion. That law of motion is, in turn, a consequence of the general relativistic conservation laws, which require that the covariant divergence of the energy-momentum tensor vanish. The field equations of the theory give the energy-momentum tensor as a function of the metric field; as early as 1920, Eddington proposed that the field equations be understood as an identification and not as a contingent relation: these equations, on Eddington's construal, specify what matter-energy *is*. When a function of the metric field which equals the momentum-energy tensor is substituted for the momentum-energy tensor in the conservation laws, the conservation laws are transformed into mathematical truths of the tensor calculus.

I surmise that, other things being equal, we prefer explanations that explain regularities in terms of necessary truths to explanations that explain those same regularities in terms of other empirical generalizations. I suppose, further, that we find particularly virtuous those explanations that explain regularities as the result of necessary truths and do so by means of connections that are themselves necessary. When such explanations succeed, they are complete, and nothing remains in need of explanation. I do not mean to suggest that any or all of these considerations exhaust the criteria by which we judge explanations. That is why there are *ceteris paribus* clauses. It was, of course, true that one of the major reasons for the inferiority of the Newtonian explanations of the perihelion advance and light deflection was that all of these explanations required hypothetical circumstances—hidden masses and concentrations of ether—for which there was no evidence whatsoever.

V

The explanatory principle provides a reason why, in each of the cases described in section II, the explanation is particularly virtuous, and why it is better than competing explanations. In some of these cases, there is even a little historical evidence that these virtues were recognized in something like the form they are given here: Kepler praised Copernican theory in contrast to Ptolemaic astronomy because the former makes the regularities of the planets a "mathematical necessity." In each of the cases considered, the explanatory power is obtained by introducing theoretical connections and structure not explicit in the regularities to be explained, and in each of these cases the connections among empirical regularities could not be made at all without additional theoretical structure. The regularity of the superior planets simply cannot be explained as the result of a mathematical truth, unless the quantities occurring in the statement of that regularity are somehow related to other quantities distinct from solar years, oppositions, and revolutions of longitude.

The same is true of Spearman's tetrad equations and of the motion of Mercury, and so on. The law of definite proportions cannot be explained as a result of the additivity of mass, unless one introduces objects other than the macroscopic samples of elements that were used in the nineteenth-century laboratory. In some of these cases, the additional structure that must be introduced to obtain the explanatory connections is structure that has turned out to be observable; in other cases, that structure has not turned out to be unobservable, or at least arguably unobservable. But observability or nonobservability has nothing to do with the explanatory virtue in question. There seem to be no grounds, save arbitrary ones, for using a principle of preference to give credence to theoretical claims when the claims turn out to be about observable features of the world, but to withhold credence when the theoretical claims obtained by the very same principle turn out to be about unobservable features of the world. The argument for atoms is as good as the argument for orbits.

I know of three objections to this argument: first, that explanation is subjective and context-dependent and therefore cannot be an objective and interpersonal ground for belief; second, that the principle of preference presented in section IV may be acceptable as a principle governing what theories we should prefer to accept or display or work with, but it cannot be acceptable as a principle governing what theories we should prefer to *believe*; and third, that the introduction of theoretical structure beyond the bounds of observation generates underdetermination without limit.

Many writers have maintained that explanation is basically a pragmatic notion. The fundamental explanatory relation is held to be something of the form 'X explains Y for person P'. Of course, what will explain Y for person P will depend on what P already knows about Y, does not know about Y, wants to know about Y, and so on. This view of explanation is surrounded by examples trading on the ambiguities of "why questions," and the context dependence of relevant replies to questions of the kind "what causes Y?" Undoubtedly, understanding is a pragmatic notion, and also, undoubtedly, we legitimately call "explanations" those replies to "why questions" (and questions of other forms) that produce understanding in an audience. None of this has any pertinence to whether or not there are objective, nonpragmatic relations between theories and statements of empirical phenomena, relations which when apprehended produce understanding and give grounds for belief. The claim that explanation is *entirely* interest-dependent has exactly as much merit as the claim that 'proof' is entirely interest-dependent: based on what one's mathematical audience knows or believes, one says some things in giving proofs, but not other things; there are arguments that meet the formal conditions for proof but which we do not call proofs because they are uninteresting (for example, because we have no reason to believe the premises, as in the "proof" of the Poincaré conjecture from the premises that (1) Glymour weighs more than 190 pounds and (2) if Glymour weighs more than 190 pounds then the Poincaré conjecture is true). An account of "proof" entirely tied to speech acts and eschewing characterization of ideal logical form would be blinding; so would a like account of explanation.

Bas van Fraassen has argued that preferences of the kind to which I have appealed in principle 1 cannot be credences or preferences as to what to believe. The reason is as follows: According to principles such as 1, Copernican theory, for example, is to be preferred to the conjunction of the empirical laws which it explains. But the conjunction of the empirical laws that Copernican theory explains is a *logical consequence* of Copernican theory. We cannot reasonably give a claim more credence than its logical consequences. Hence, the preferences in principle 1 cannot be credences or preferences for belief.

Van Fraassen is right, so far as this goes, but the preferences may, nonetheless, be tied to belief. It can be irrational to believe a collection of empirical laws, to believe that a particular theory entails these regularities, is tested by them, and explains them better than any other possible theory, and still not believe the theory. The irrationality is akin to that of believing certain premises, believing that some other sentence is a logical consequence of these premises, and refusing nonetheless to believe the further sentence. There is inductive closure quite as much as deductive closure. The preference stated in principle 1 is a preference for a certain candidate for the inductive closure of the regularities explained. If theory T is preferable to theory Q, according to principle 1, that is not to be understood to mean that T is to be believed rather than Q, or that there is reason to believe T rather than Q, for Q may be a logical consequence of T. Rather, the preference is to be understood to mean that the inductive closure of the laws explained is (the deductive closure of) theory T rather than (the deductive closure of) theory Q. More simply but less exactly, the preference is for believing T rather than believing *merely* Q, when T and Q are consistent, or for believing T rather than Q when T and Q are not consistent.

Some years ago, Quine argued that once we introduce language which does not describe observable features of the world, underdetermination is rampant. "Rampant" means infinite here, for if only a finite number of theories are consistent with all possible evidence of some kind and satisfy our inductive canons, then there is a preferred conclusion, namely their disjunction. Such sweeping views are unconvincing, for they lack both a convincing account of theoretical equivalence and a plausible characterization of inductive canons. In more restricted cases, where a characterization of the class of possible theories is available and clear and applicable criteria for sameness of theories are at hand, familiar arguments for underdetermination fail when inductive canons are applied. For example, H. Reichenbach's well-known arguments for the underdetermination of space-time geometry fail when the inductive canons I prefer are applied to space-time theories.

It is certainly the case that the principle of explanation that I have described, even with elaborations about necessity and identity, is insufficient as a foundation for inductive logic, and if that principle were the whole of the matter, underdetermination would indeed be rampant. Nonetheless, principle 1 in conjunction with other available inductive principles, can be extremely powerful, and can in some contexts virtually eliminate underdetermination—or so I conjecture. Consider the context provided by Spearman's kind of theory, where the data consist of correlations among some determinate set of variables. The theories consist of any set of digraphs without cycles, connecting the measured variables with one another or with unmeasured variables, and the associated linear equations. Under the assumptions about error made before the equation, they can be reduced to equations relating the "path coefficients" to one another and to the measured correlations. The correlations may satisfy equations such as Spearman's tetrad equations, or analogous equations involving triples of correlation coefficients, quadruples of correlation coefficients, and so on. Given a set of data satisfying a determinate set of equations of this kind, of the infinity of digraphs consistent with this data, a proper subset will reduce to mathematical truths all and only the equations of the set satisfied by the data. This set of digraphs will be those preferred by principle 1.

In some cases the set of preferred digraphs will, I believe, be finite and small. For example, in addition to the digraph illustrated in connection with Spearman's explanation of his correlation equations, I know of only three kinds of directed graphs that satisfy the principle for the three tetrad equations Spearman assumes holds empirically. They are

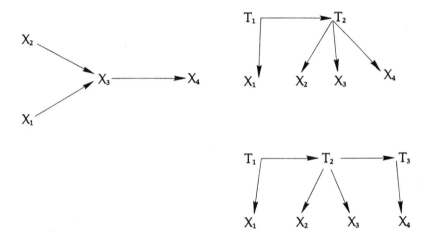

The first digraph, which contains no unmeasured variables, really stands for an entire set of digraphs, since any two variables can be interchanged in the digraph without affecting the derivation of the tetrad equations. Any of the digraphs of this kind can be distinguished from those which introduce latent variables, since digraphs of this kind imply restrictions on partial correlations that are not implied by digraphs with latent variables. Various other empirical constraints may distinguish the three alternatives with unmeasured variables from one another in the context of a larger theory. Even when other empirical constraints are not available, other methodological constraints may be. In *Theory and Evidence*, I described a strategy for testing systems of algebraic equations in real variables with known coefficients. In the context of linear theories of the kind under consideration here, principles founded on bootstrap testing may not themselves lead us to prefer theories with unmeasured or latent variables, but they may lead to determinate preferences among a set of theories all having latent variables. Starting with Spearman's supposed tetrad equations, we have four classes of theoretical digraphs that satisfy principle 1. Partial correlations can either establish or eliminate the class of digraphs containing only measured variables. Of the remaining three digraphs, Spearman's original model is the one preferred on bootstrap principles in the absence of other information. That is because the other two digraphs implicitly contain theoretical coefficients which cannot be evaluated from measured correlations even assuming the correctness of the causal relations they postulate. In the econometricians' jargon, they are not "identifiable."

This is only an example, and a conjectural example at that. I mean it to illustrate nothing more than that there may well be many surprises in store when issues of underdetermination are examined in more modest and more detailed ways. Certainly, there is at present no particular reason to believe that underdetermination is rampant once we move beyond the observable. Underdetermination is, I expect, indestructible in another way, for logic does not determine a unique set of inductive canons, perhaps not any set of inductive canons. Thus, antirealists may still have recourse to denying principles of preference or inference such as those I have tried to describe here. My point has not only been to give that description, but to argue that the principles I have described are principles we use in our sciences to draw conclusions about the observable as well as about the unobservable. If such principles are abandoned *tout court*, the result will not be a simple scientific antirealism about the unobservable; it will be instead a not so simple skepticism.

NOTES

1. J. Dumas, *Lecons de Philosophie Chemique* (Paris: Gauthier Villars, 1937), 178-179. This edition is a reprinting (with slight alteration of title) of the 1837 edition.
2. C. Hempel, "The Theoretician's Dilemma," in *Aspects of Scientific Explanation* (Glencoe: Free Press, 1965).
3. Cf. P. Duhem, *The Aim and Structure of Physical Theory* (New York: Atheneum, 1974), and W. V. O. Quine, "Two Dogmas of Empiricism," in *From A Logical Point of View* (Cambridge: Harvard University Press, 1953).
4. C. Spearman, "General Intelligence Objectively Determined and Measured," *American Journal of Psychology* 15 (1904): 201-293.
5. In the first class fall the theories of Hempel and Oppenheim, D. Kaplan, J. Kim, and more recently, J. Thorpe, B. Cupples, and others. In the second fall theories such as those of Salmon, Brody, Friedman, and Causey, and the third class includes theories such as those of van Fraassen, B. Skyrms (I believe), P. Achinstein and, recently, Putnam.

6. C. Hempel and P. Oppenheim, "Studies in the Logic of Explanation," in *Aspects of Scientific Explanation.*

7. Cf. R. Eberle, D. Kaplan, and R. Montague, "Hempel and Oppenheim on Explanation," *Philosophy of Science* 28 (1961): 418-428.

8. D. Kaplan, "Explanation Revisited," *Philosophy of Science* 28 (1961): 429-436.

9. Cf. W. Salmon, *Statistical Explanation and Statistical Relevance* (Pittsburgh: University of Pittsburgh Press, 1971).

10. M. Friedman, "Explanation and Scientific Understanding," *Journal of Philosophy* 72 (1974): 5-19.

11. R. Causey, *Unity of Science* (Dordrecht: D. Reidel, 1979).

12. B. Brody, *Identity and Essence* (Princeton: Princeton University Press, 1980).

13. I am indebted to my student, Jeanne Kim, for suggesting this example to me.

10

Truth and Scientific Progress

Jarrett Leplin

Philosophers of science are increasingly taken with the following apparent paradox: theories that excelled under criteria now employed in evaluating theories ultimately proved unacceptable; therefore, it is likely that the best current theories, and even better ones that we might imagine overcoming what defects we now recognize in current theories, will prove unacceptable. Thus, we have a strong inductive argument against the ultimate acceptability of theories that we have strong inductive grounds to accept.[1] It is common to summarize this situation by some such equally paradoxical remark as: "We really know that all of our scientific beliefs are false."

Of course the "paradox" is not genuine. Eschewing induction blocks inference to the fate of current theories from the record of past theories. But then, acceptance of any particular theory on the basis of evidence bearing individually on it is equally blocked, and the negative, general implication as to the prospects for scientific knowledge is sustained. Philosophers of science on many fronts are understandably in retreat from the traditional view that the growth of science delivers ever closer approximations to fundamental truths about the physical world.

Part I of the present paper examines some examples of this movement so as to establish a general perspective on the problem of connecting truth with progressive theory change. Part II defends the traditional view that progress attests to truth by developing the thesis that science exhibits certain forms of progress for which realism with respect to scientific theories is the required explanation. In Part III the argument of Part II makes a troubled peace with the skeptical induction just indicated. The scope and limits of the realism that emerges are then delineated.

I

A number of diverse philosophical views associated with Thomas Kuhn may be brought to bear on the problem of connecting truth with scientific progress. Here is a perspective on Kuhn. He begins as a scientist struck by the superficiality and outright ignorance of scientists as to the historical background of their current ideas. He invents an interpretation of science which explains such failings. The explanans is incommensurability: intellectual traditions are sufficiently circumscribable and independent that understanding one's intellectual heritage has no part in the mastery of current knowledge, but is solely the task of the historian. Thus, incommensurability emerges as the cardinal thesis of Kuhn's philosophy of science. Primarily on the basis of this thesis, Kuhn is widely read as denying any connection between truth and progress. In fact, he denies explicitly that any sense can be found in which a theory better approximates the truth than does the theory it replaces.[2] Whether Kuhn merely despairs of finding such a sense, or contends that incommensurability precludes there being one, is unclear. But certainly, on Kuhn's view, the increases of predictive accuracy, scope, and fertility of theories in which progress consists do not attest to the truth or approach to truth of hypotheses. The consequences of hypotheses may be judged true when sustained by experiment; but such judgment is intratheoretic, inaccessible to proponents of alternative theories.

Imre Lakatos, for all his outrage at this last bit of relativism, is equally explicit in divorcing progress from truth. Progress is a matter of increasing Popperian verisimilitude—the excess of corroborated over falsified consequences of theories. But, whereas Popper is prepared, on the strength of Tarski's definition of truth for formal languages, to interpret increasing verisimilitude as approximation to truth in the classical sense of correspondence with objective facts, Lakatos rejects this interpretation as a "dangerously vague and metaphysical idea."[3] The danger, evidently, is that we are led by Popper's view to suppose the ultimate constituents of the world to be more like the items purportedly referred to by current theory than like those purportedly referred to by past theory; more like fields, say, than like matter and force. And there is no intuitive basis for this assessment. But Lakatos seems simply to have confused the view that the classical correspondence theory of truth is the intuitive or preanalytic idea of truth and of the goal of science, with the view that scientific advance toward truth in this sense delivers ever more intuitively plausible theories.[4] Popper was certainly not endorsing the latter view in supposing increases of verisimilitude in his technical sense to indicate increases of classical verisimilitude. Popper, as everyone knows, believes that scientific progress decreases the likelihood or plausibility of theories by decreasing their probabilities so interpreted.

Lakatos accepts a Platonized version of Popper's third world; the growth of scientific knowledge is an unarguable datum for Lakatos.[5] But this position fails to connect scientific knowledge with truth. A connection between knowledge and truth is characteristic of analyses of second world knowledge, where the truth of a belief is required for attribution of knowledge to the believer. World three exhibits no such discrimination. To suppose that it should is to confuse objective knowledge with certain knowledge which, if it exists in world three at all, is confined to purely deductive relations within mathematical systems. All theories, at least all developed, articulated theoretical systems to emerge in the course of human history, are admitted undistinguished as to truth value. Indeed, the hallmark of citizenship in world three is autonomy, meaning roughly that these are things that can initiate influences on world two, that we can learn from. And what we learn from primarily, according to Popper, are mistakes.

Crucial to Popper's treatment of rational theory replacement as increasing approximation to truth is the thesis that rationally replaced theories are falsified. Lakatos rejects this thesis because of the underdetermination by experience of singular statements formulating predictions by which theories are tested. The selection of research programs must inherit the conventional freedom attaching to the selection of such statements. Perhaps this result underlies Lakatos's ironically Kuhnian view that the progressiveness of a research program reflects more on the talents of its practitioners than on its scientific merits. If, as Lakatos evidently believes,[6] the more absurd of competing research programs may, with imaginative and skillful patronage, emerge the more successful in anticipating novel facts, then Lakatosian progress is clearly weaker than Popperian. Scientists concerned to advance knowledge are deterred from *ad hominem* rejection of rival research programs only by vanity. Assessments of the rationality of theory choice are not merely irremediably retrospective for Lakatos; they are irremediably fallible. For such assessments commit us to undecidable counterfactuals to the effect that degenerating programs would not have proven more progressive than their preferred rivals if pursued. Lakatosian progress cannot be said to approach truth because it cannot be said to replace error. Rejection is no more falsification for research programs than for the individual theories replaced in their construction.

Larry Laudan's *Progress and Its Problems* is probably the most explicit recent rejection of the relevance of truth to the analysis of scientific progress.[7] For even knowledge, which might not, after all, prove so readily dissociated from truth as Lakatos imagines, finds no role in Laudan's approach. Laudan neither denies that scientific statements have truth values nor rejects the possibility of a scientific theory being true. But he infers from past philosophical failures that no sense can be made of the notion of relative approximation to truth, and that in the inductively unlikely event that any theory is true this fact about it would be in principle unknowable. To suppose truth to be a goal of science, then, portrays the activity of science as irrational, presumably on the principle that rational behavior has an end whose achievement is at least recognizable. But it is a condition of adequacy for any theory of rationality that our preanalytic intuitions as to the rationality of paradigm cases of progressive theory change be sustained. Therefore, truth is not a goal of science, and individual scientists whose motivations Laudan acknowledges to include a quest for truth are presumably confused in some way.[8] The task of Laudan's work is to give an analysis of progress, in terms of which rationality can in turn be defined, which makes no use of the concepts of truth and knowledge.

This task, if achievable, would certainly constitute an interesting and worthwhile alternative to traditional attempts to understand the rationality of scientific change. There are, however, a number of reasons for skepticism. One is that the record of philosophical failure, which so impresses Laudan as to induce him to discount the very possibility of understanding progress in terms of truth, includes the career of instrumentalism from which Laudan's own problem-solving model of progress is insufficiently distinguished. The claim that scientific statements possess determinate truth values amounts to little if coupled with the claim that such truth values are in principle indeterminable nor even subject to probabilistic estimation.[9] Laudan does maintain, in apparent opposition to instrumentalism, that research traditions impose certain "ontological commitments" on their constituent theories. But the nature of these commitments is unclear, as none of Laudan's criteria for the appraisal of theories are criteria for the acceptability of their ontological claims. Another reason is that Laudan's own problem-oriented approach to his task creates its own philosophical problems which it has yet to show progress in solving. On many crucial points, including the proper understanding of what constitutes a problem or a solution and, in particular, the grounds for evaluation of the latter, we have only promissory notes.

But I shall not pursue such criticisms here. What I stress instead is the counterintuitiveness of an account of scientific progress that finds no role for truth or knowledge. That science does progress and that scientific change is, in large measure, rational, are presuppositions of Laudan's work. By what reasoning does he hold the claim of science to deliver knowledge of the physical world any less entitled to presuppositional status? Philosophers prepared to assume the rationality of scientific change, and to reject interpretations of science that depict theory choice as irrational or arational on this account alone, have invariably credited science with the status of a knowledge-acquiring enterprise. Laudan's dissociation of these assumptions is a radical step requiring some compelling motivation. Until he provides one, or salvages scientific knowledge through some new, equally radical analysis of knowledge requiring neither the truth nor the likelihood of successful knowledge claims, his work is seriously incomplete.

An apparent exception to the movement away from understanding scientific progress as truth approximation is Dudley Shapere.[10] Shapere holds that truth can be salvaged as an achievement of science if truth as a metaphysical ideal is reduced to epistemology. The truth of a scientific statement amounts, for Shapere, to the absence, after a reasonable period of critical inquiry, of specific reasons for doubt with regard to the statement. Thus if a statement thrives according to the evaluative standards of science, it qualifies as true despite the possibility, admittedly ever present, that specific reasons for doubt will yet emerge, and despite the strict consistency of its denial with the corpus of scientific evidence. A statement is to qualify as true if research appropriate by the standards and methods of current science fails to reveal specific grounds for doubting it. But, of course, such grounds might subsequently emerge. Moreover, standards and methods might subsequently so change that the evidential basis of the statement comes to be regarded as inadequate or inappropriate.

According to Shapere, however, it is consistent simultaneously to maintain that the statement is true and to acknowledge these possibilities, to acknowledge even the possibility that such changes will be rational and progressive so that the original endorsement of the statement will subsequently be regarded justifiably as the unfortunate error of a weak or misguided methodology. We might go so far as to imagine the statement falling into such disfavor as to impugn the scientific standards employed in endorsing it. In Shapere's view, the statement qualifies as true despite all such possibilities.

Clearly, various metaphysical features of the concept of truth—for example, that if a statement is true it remains so whatever the vicissitudes of belief, and that a statement might be false even if there are good reasons to believe it and no good reasons to disbelieve it—are being rejected here. What is the justification for this? Does this not amount simply to redefining 'truth' so as to get the result that science achieves truth? Is Shapere's approach any less a retreat from truth than the approaches of philosophers who dissociate truth from knowledge?

Such cogency as Shapere's position has emerges from its response to these challenges. What Shapere ultimately is arguing for is the abandonment of an abstract metaphysical ideal of truth which has no connection with the intellectual processes by which knowledge is acquired and which is, accordingly, unrealizable in principle. Shapere has long insisted that such concepts as "observation" and "evidence" be so analyzed as to reflect their actual use in scientific reasoning, rather than imposed as the abstract philosophical tools of a presuppositionist philosophy of science. By this he means that situations actually regarded as observational or evidential by scientists in the articulation of successful theories are to be accorded such status by philosophical theories of observation and evidence. His move now is to treat truth and knowledge in the same way. A criterion of acceptability for theories of truth and knowledge is that they admit paradigm cases of successful science. A truly empiricist philosophy of science must reject presuppositionism altogether, deferring to science itself the identification of what is true and what is known.

But there is a conceptual difficulty in so sweeping a rejection of presuppositionism. There must be presuppositions at some level if there are to be standards against which the legitimacy of any particular presupposition is assessed. Shapere apparently contests this, maintaining that standards at any level may be rationally violated as science changes. Yet Shapere himself is not offering a presupposition-free philosophy of science in place of the tradition of measuring science against presupposed metaphysical ideals. He is offering a different set of presuppositions on which a particular view of the status of scientific results constrains what we are allowed to mean by 'knowledge' and 'truth'. Successful science explicates the concepts of knowledge and truth, and for criteria of success we simply defer to consensus within science itself. The disadvantage of this substitution of presuppositions is that it abandons the very idea that the scope and limits of the scientific enterprise as a whole are subject to philosophical assessment.

An interesting novelty of Shapere's approach is its emphasis on the continuity and stability of scientific results. While recognizing the instability of scientific methods and standards, and rejecting, on that account, presuppositionist philosophies of science, Shapere requires considerable survival power of those statements that are to qualify as true. For the fact that specific reasons for doubt do not readily appear upon the introduction of a scientific hypothesis will hardly be enough to render it true. Shapere no more specifies the amount of critical inquiry that a hypothesis must survive to be true than Lakatos imposes a time limit for the rejection of degenerating research programs or than Kuhn specifies the number of anomalies required for revolution. But Shapere invites us to imagine ourselves in possession of a comprehensive theory to which no compelling objections arise over a period of hundreds or thousands of years, and he intimates that reluctance to acknowledge the truth of such a theory could only be philosophical perverseness.

Polemics aside, the interesting aspect of this picture is its utter disregard of the inductive argument against any such foreseeable stability of scientific theory, whose inducement to skepticism with respect to truth I began by noting. If Shapere's approach is to yield its desired result that science achieves truth, there must be some historical counterargument suggesting that science does or will produce theories whose rejection ought not to be anticipated. We will need some account of continuity in science at the level of theory, whereas any account of continuity consistent with the skeptical inductive argument would appear limited to the empirical level at which theories are tested. Laudan, for example, holds that it is only the base of empirical problems that exhibits continuity in science; their solutions are short-lived.

Just such a historical counterargument as Shapere requires has coincidently been attempted by Levin.[11] It depends, of course, on assuming what Laudan is at pains to deny, namely, that theory change is cumulative. Levin claims that each successive theory is obliged to explain all the facts that its predecessor explained, and more. It must also explain some facts that its predecessor failed to explain, in particular, those responsible for the rejection of its predecessor, and it must explain why its predecessor, although false, succeeded where it did. Given these conditions of adequacy for a new theory, Levin maintains that theory replacement is an ever more arduous activity. The greater the number of facts to be explained, the more difficult it is to explain them. So the chances of coming up with a new theory are reduced in proportion to the length of the historical succession it is to join. Every scientific revolution, suggests Levin, makes it less likely that another will occur.

This turns the skeptical inductive argument on its head; instead of inferring that present theory will be rejected from the fact that past theories were rejected, we are to infer that present theory will be retained from the fact that past theories were rejected! Presumably, were history to reveal few rejections and great longevity for current theory, then revolution should be anticipated because it would be so much easier to accomplish!

One reply to this remarkable argument which I wish to reject is a denial of cumulativity. It is well established that successive theories typically address different problems or seek to answer different questions, in part because some of the questions a theory is responsible for answering receive their first formulation in the context of the theory itself. Perhaps no theory explains all the facts thought to be explained successfully by its predecessor. But this does not refute cumulativity. Cumulativity I take to be the thesis that facts established by theories continue to be recognized as facts from the vantage point of future science, so that once a fact has been established, its status as a fact will continue to be inferable from future theories. This thesis is unimpaired by the finding that what were formerly considered established facts are no longer so regarded, provided current theory rationalizes their rejection. Nor does appropriation of established facts by theories other than those which replace theories first establishing them impair cumulativity. Science as a whole may exhibit cumulativity if individual theory transitions do not. What would refute cumulativity are the possibilities that in the course of scientific growth, no facts are established at all, for then there is nothing to accumulate, and that purportedly established facts are neither inferable nor deniable on the basis of future theory. Examples of change in the problem situations confronting successive theories in themselves realize neither possibility.[12]

There are, however, more telling replies. Levin's argument suffers from an exclusive focus on the process of theorizing. As the scope and depth of theories increase, so does the sophistication of experimental procedures, both as to the precision and range of measurements and as to the ability to produce technologically new types of phenomena requiring theoretical understanding. As theories become more comprehensive, the ability to test them, and with it the likelihood of producing disconfirmations, increases. So although the conditions of adequacy for a replacement theory are more severe than those for current theory, the ultimate acceptability of current theory is less likely than was that of past theory. Certainly, Levin's argument will fail to impress a Popperian who believes that the more a theory explains, the greater its content, the easier it is to falsify.

The real issue here is the connection between disconfirmation and rejection or, more clearly but less precisely, between falsification and theory replacement. Levin presumes, as do many philosophers, that no theory, however unsatisfactory, can rationally be rejected unless a superior theory is available to replace it. If it is possible to make an adequate case against a theory in the absence of an alternative, then Levin's argument surely fails. For the difficulties of constructing a new theory may then be irrelevant to the survival of present theory. I have never understood the arguments which are supposed to show that the absence of a better theory in itself guarantees the continuing acceptability of whatever theory is available. Most such arguments are nothing more than complications of historical examples which are equally well interpreted as evidence of the ability of scientists to come up with a new theory when one is wanting. Perhaps the clearest attempt at such an argument is Lakatos's, which depends on assuming that no amount of empirical or methodological difficulty (nor, for that matter, inadequacy relative to alternatives) ever legitimates the conclusion that any theory is false. On this view, so much mainstream scientific judgment becomes illegitimate that one wonders how Lakatos can make maximization of the scope of internal history a desiradum for historiography of science.

It is worth adding that if we allow that theory rejection requires theory replacement, Levin's argument is then vulnerable to the conjecture that the ability of scientists to construct new theories improves with experience. Levin's example of the cavemen with five facts to explain suggests that we are to regard the faculty for theory invention as a diachronically fixed, innate endowment incapable of evolution. But we are not simply cavemen with more than five facts to explain, who, therefore, face a tougher task. We have learned how to construct theories and how to distinguish promising from unpromising directions of scientific creativity.[13] Analysis of the epistemic basis of these abilities reveals the role of experience. Levin seems to have been taken in by the old view that the creative aspects of science are not a fit subject for comprehension.

Although flawed, Levin's optimistic reversal of the skeptical inductive argument is instructive. For its irony mirrors an irony of the original argument by which the core issue for a theory connecting truth with scientific progress may be grasped. I intimated that Levin is committed to the converse of his position, to the view that were history to reveal great longevity for theories and little in the way of theory replacement, then future replacement would become more likely. The reason is that were Levin to maintain that under such hypothesized conditions replacement remains unlikely, we would get the result that history is simply irrelevant to the prospects of current science. I consider this result inadmissible, for while the fate of a theory might not depend on the availability of a successor, it surely does depend on the conditions of its provenance.

As many philosophers have argued, there are dimensions of theory appraisal to which the projection of a theory against its background of methodological and metaphysical commitments and their evolution is crucial. Now, if we consider the converse of the original skeptical argument, we get the result that were past theory to remain viable and not require rejection, then current theory also should be expected to survive.

But this inference is unacceptable as we are dealing with exclusive theories among which a decision is ultimately necessary. The original argument therefore suggests that current theory is in trouble whatever the fate of past theories. *What turns out to cause epistemic problems for current theory is not so much the fact that past theories were rejected as the fact that there were past theories at all, which are inconsistent with current theory.* Current theory is unreliable because history has produced alternatives.

To put the point another way, consider the fact that a necessary condition for theory succession to approach truth is that replaced theories be wrong. For surely, if they are not wrong, then we do not approach truth by replacing them. It is partly his conviction that this condition cannot be known to be satisfied that leads Lakatos to reject any connection between progress and classical verisimilitude. Now, it is ironic that the *satisfaction* of this condition hypothesized by the original inductive argument leads to a skeptical conclusion. The argument in effect maintains that theory change does not approach truth whether or not this condition is satisfied. The real thesis appears to be that *truth is not something that can be approached.* It does not come in degrees. It must be achieved all at once in a theory that never gets abandoned, or not at all. Only such an immortal theory which obviates history can have any connection with classical truth. Why is Shapere so interested in a theory that lasts thousands of years if it is not that for such a theory history no longer counts? The vicissitudes of attempts, failures, and new attempts which prepared its introduction become prehistory or protoscience of no possible relevance to its epistemic credentials. The argument from actual history to the unreliability of current theory really argues that no theory is acceptable so long as it has a history to be reckoned with.

The only way I know to counter this view, the only way to salvage a role for truth in the analysis of scientific progress, is to take the position that there are degrees of truth, that truth can be approached. Even if no theory ever to be produced is completely true, later theories are, for the most part and on balance, truer than earlier ones. This is the view I shall seek to defend.

II

My approach will have much in common with views advocated or once advocated by Hilary Putnam. Putnam has espoused realism with respect to scientific theories and claims to have shown that only the classical conception of truth offers an adequate account of the nature of science. I shall begin by discussing Putnam's argument.[14]

Putnam maintains that scientific progress, at the level of increasing predictive success, is an established fact. He offers realism as an explanation of this fact. In particular, that the purportedly referential terms of mature science refer and that the laws of mature science are approximately true are *hypotheses* which render this fact comprehensible. If we reject these realist hypotheses, the success of science becomes incomprehensible—a "miracle," says Putnam.

There is an embellishment to the effect that the acceptance of these hypotheses by scientists renders their behavior comprehensible. Scientists typically seek new theories whose laws reduce to those of established theories "in the limit" and whose terms are coreferential with those of established theories. It would be perverse of them to accept such constraints on theorizing were they not realists in the sense of these hypotheses. This embellishment is problematic, because if one asks why this would be perverse, the answer is that violation of these constraints makes it easier to meet the empirical requirements for new theories. But if this is so, why does empirical progress require realism for an explanation? Empirically successful methods free of realist constraints would themselves provide the explanation. The point of the embellishment must really be that realist constraints are, rather than a hindrance to theorizing, a help in selecting among possible theories those most likely to be empirically successful. That they have this effect is again an explanandum for the realist hypotheses.

One might attempt to counter Putnam's argument with a version of the skeptical inductive argument. We cannot explain the success of past theories by invoking the truth or reference of past theories, as we do not believe that past theories are true or referential. Therefore, it is unlikely that the success of current theory can be explained by invoking these attributes. That the realist hypotheses do explain current success is a crucial premise for Putnam, however. For the claim that nothing other than realism explains scientific success is by itself an insufficient case for realism. Whatever force this argument has against Putnam depends on its suppressed inference from the failure of truth and reference in past theories to their failure in current theory. Only this inference provides the argument's challenge to the current explanatory status of realism. So it is really the original skeptical argument that must, once again, be confronted. Putnam acknowledges this by allowing that his reasoning depends on blocking the skeptical induction. But he provides no means of doing this.[15] One might, therefore, simply rest the case against Putnam.

This, in effect, is what M. Hesse does.[16] But the matter is not quite so straightforward. Hesse appears to allow that the realist hypotheses do explain the empirical success of science but contends that the skeptical induction, which she formulates as a principle of "no privilege" for current science, requires that some other explanation short of realism be found. Thus, the target is actually Putnam's second premise, the claim that there is no nonrealist explanation. Unfortunately, Hesse fails to produce one. What she produces instead is a "reconciliation" of her "no privilege" principle with the "principle of growth," that "science does exhibit apparent cumulative predictive success." The reconciliation consists in showing that sentences at an observational level, including some formerly theoretical sentences which made no purported references to hypothetical entities and were therefore able to shift over to the observational level, survive scientific revolutions. But what was to be explained was not simply how it is possible for science, at periods separated by revolution, to agree on some basic empirical facts—there are nearly as many ways of doing this as there are critics of incommensurability. Rather, the explanandum is the dramatically successful predictive power of theories. The explanandum, one might say, is the achievements themselves, not the conditions of their accumulation. If theorizing does not approximate truth, if the entities it postulates do not exist and their descriptions are not approximately satisfied by anything that does exist, why are its predictions so successful?

Laudan's reply to Putnam also provides no explanation. The version of the skeptical inductive argument on which Laudan relies is the claim that the inference to the truth or reference of theories from their predictive success is invalidated by history. Thus, Putnam is wrong to cite predictive success in support of realism. If the realist hypotheses really are empirical, as Putnam says they are, then they must be regarded as falsified by historical evidence. The explanatory status of the hypotheses has no role in this criticism. This is appropriate enough, as Laudan is hardly in a position to deprive any hypothesis of explanatory status on grounds of epistemic liability. But the omission results in a misrepresentation of Putnam's reasoning. Putnam does not infer the truth of realism from the predictive success of science. Rather, he infers the truth of realism from its alleged, unique ability to explain the predictive success of science. Once again, the telling criticism seems to be one that attacks realism directly on the basis of a skeptical reading of history, and then concludes indirectly either that realism fails to explain predictive success because it is false, or that the explanation it provides is unacceptable. The explanatory problem Putnam has raised remains outstanding.

The general result seems to be this: the denial of realism is compatible with a certain form of progress, perhaps even with growth of knowledge in some as yet inadequately articulated third world sense. Moreover, the form of progress that science exhibits seems an insufficient evidential basis for realism, as such progress has been achieved by discredited theories. Nevertheless, progress seems to require realism for an explanation. Specifically, the fact that not only does science exhibit an accumulation of knowledge at the pragmatic, empirical level but theory transitions produce marked *increase* of predictive success, seems explainable only by the superiority, according to realist standards, of theories over their predecessors.[17]

The superiority required is not primarily referential. New theories may purport to refer to more, to fewer, or to different entities than their predecessors, but that is not a condition of their superiority. Moreover, if their replacement in turn required referential failure, an initial judgment of their referential superiority could not be sustained. What is required referentially is stability; new theories at least sometimes preserve the references of their predecessors.[18] Their superiority must primarily be a matter of truth; a new theory of greater predictive success, if it does not introduce new references which in turn are preserved, must more closely approximate truths about the nature of the referents of the replaced theory. Short of recognizing superiority of this sort, we must either regard a major form of scientific progress as unexplainable or we must deny that progress of this form actually occurs. The latter position leads directly to historical relativism; it is, in effect, Kuhn's view that each major theory determines its own domain of application, so that Aristotelian scientists can be credited with having learned as much or more about their world as Newtonian scientists learned about theirs.

The apparent relationship between increasing predictive success and the hypothesis of truth approximation, whereby the latter is needed to explain but is inadequately supported by the former, is typical of science itself. Galileo was unable to explain the tides because his aversion to astrology prevented him from recognizing celestial influences. Newton explained the tides by celestial influences. Of course, the phenomenon of tidal motion in itself is insufficient reason to accept Newtonian gravitation even if no other explanation seems possible. But its combination with additional otherwise unexplained and unrelated motions is compelling. When an explanans is underdetermined by but uniquely explanatory of its explanandum, one investigates its explanatory potential in other areas, particularly areas that have also proven resistant to explanation and that differ significantly from the explanandum.

The hypothesis of truth approximation has such potential, for there are established forms of scientific progress distinct from increasing predictive success. There is, most obviously, increasing predictive power, the fact that a greater number and diversity of predictions are generated by theory change, which there is certainly no a priori reason to expect to accompany or to be accompanied by increasing success (unless, with

Lakatos, one reckons success without regard to failure). More generally, there is increasing explanatory, problem-solving, and question-answering power as science develops. And there is a form of progress that consists in extension of the scope of observation, in the ability to observe newly postulated entities. Especially important is the ability to observe such entities directly, where 'direct observation' has not its philosophic sense of observation free of theoretical presuppositions, but its scientific sense of achieving the best possible access or evidential situation with respect to the entities which the theory of them allows. I shall argue that truth approximation is vital to the explanation of these phenomena as well as to the explanation of increasing predictive success, so that combination of explananda will increase the credibility of realism.

Of course, it is possible to question these alleged explananda, just as the presumption of increasing predictive success has been challenged. Moreover, while increasing predictive success may seem too manifest an accomplishment to be denied, the other forms of progress claimed for science are sufficiently abstract and complex to issue in such technical difficulties as invite skepticism as to the very claim that such progress occurs. It is important to recognize that rejection of this claim, as much as that of increasing predictive success, introduces relativism. It asserts, in effect, the symmetry of judgments as to the capacities of successive theories when such judgments are relativized to the theories themselves. Each theory fares better by its own standards, and there are no transtheoretic standards to which to appeal. Thus, A. Grünbaum contests Popper's contention that general relativity (E) answers more questions than the Newtonian theory of gravitation (N) by denying, implicitly, that there are transtheoretic criteria for the legitimacy of questions. E answers more Einsteinian questions; N, more Newtonian ones.[19]

Discussion of this example will serve both to defend the claim that scientific change increases problem-solving and explanatory power, and to develop the role of truth approximation in explaining these phenomena. N and E are incompatible theories. Therefore, a nonmetrical (qualitative) comparison of their Tarskian logical content or Popperian information or empirical content under the set inclusion relation will not yield the result that the transition from N to E increases verisimilitude. Nevertheless, Popper maintains that a nonmetrical comparison can sustain the intuition that E exceeds N in explanatory power, by focusing initially on questions which E and N can answer. For E answers, with at least equal precision, all the questions N answers, and more. Grünbaum produces as counterexamples a series of questions which N answers with precision but which cannot be formulated in E. Each question has a presupposition, either permissible or realized according to N, which contradicts E. And it is clear that there is no end to the generation of such questions, precisely because E and N are incompatible. As an initial example, consider the question:

(1) Why is the orbit of a planet of negligible mass subject to the gravitational attraction of the sun alone Keplerian?

N answers (1) by straightforward mathematical deduction from basic laws. E denies the presupposition that the orbit is Keplerian, implying to the contrary that the planet precesses.

Now despite Grünbaum's argument, one is inclined to suppose that there is an intuitive sense in which E does answer all the questions N answers and more. This is the sense in which 'question' is taken to mean "question whose presuppositions are consistent with current theory." The intuitive reply to Grünbaum, then, is that E answers all the questions N answers which satisfy this restriction, and that *it is legitimate to impose this restriction*. But how can the latter claim be defended? If the issue to be decided were which of the theories E and N is the better theory, and the question-answering power of E and N were being compared with a view to informing this decision, then there could be no defense. The imposition of Einsteinian standards in assessing the legitimacy of questions would be circular. But this is not the issue, at least not here. We start with the fact that E is better, a fact established by E's comparative predictive success. The issue is how this fact is to be explained. And we find that this fact enables us to recognize a second fact also requiring explanation, that E exceeds N in question-answering power.

This reply, as it stands, will not do. From the fact that E exceeds N in predictive success, it by no means follows that E is superior to N in such respects as would justify restricting legitimate questions to those whose presuppositions conform to E. The very point at issue is whether E exhibits additional relative merits that can serve as explananda for its relative truth. This issue is prejudged by assuming E-presuppositions superior to N-presuppositions. Nevertheless, the independence of the assumption of E's superiority with respect to predictive success from the circularity of the reply is relevant to the legitimacy of the intuition that E exceeds N in question-answering power. Consider the question:

(2) Why does the perihelion of Mercury precess?

Although N answers (1), it can also answer (2). For N is not committed to the satisfaction by Mercury of the conditions hypothesized in (1). E also answers (2), differently of course. Thus, the presupposition of (2) would seem compatible with both N and E. Does this indicate a transtheoretic standard of legitimacy for questions? It depends on how (2) arises. Its presupposition might be deduced from each theory independently as the idealizing conditions hypothesized in (1) are corrected, so that each theory judges (2) legitimate by its own standards. But what if the deduction is not performed or the idealizing conditions are not yet correctable, and (2) arises *empirically*? Must we then await these theoretical developments for a decision as to (2)'s legitimacy? Consider

> (3) Why does the perihelion of Mercury precess by 5,650 seconds of arc per century?

Presumably the presupposition of (3) is not obtainable deductively from N. Yet neither does it contradict N in the way that the presupposition of (1) contradicts E. Is (3) legitimate for N? If so, can N answer (3)?

It seems to me that the answers to these questions are "yes" and "no," respectively. (3)'s legitimacy for N was acknowledged by proponents of N who therefore tried to answer (3), but failed. Their failure indicates that (3)'s presupposition is incompatible with N. But this does not render (3) dismissible as illegitimate, for the simple reason that (3) arose as an empirical question and is, therefore, legitimate transtheoretically. None of Grünbaum's examples are empirical questions, but if we consider such questions we destroy at once the apparent symmetry of question-answering power of competing theories. For E answers all the questions legitimate by its own or by transtheoretic standards that N answers, and more. But N fails to answer all the questions legitimate by its own or by transtheoretic standards that E answers, for it fails to answer (3). If we can assume that N does not exceed E in scope of application, so that E's superiority in question-answering power is not purchased at the expense of legitimating fewer questions, then we can infer from the destruction of this symmetry that the E to N transition increases the number of questions answered.[20]

Now consider the following reply. N can answer (3). There are auxiliary hypotheses consistent with N that enable N to yield precisely the empirically determined amount of Mercury's precession. The problem is not that N cannot answer (3); the problem is that N's answer is incorrect or unsubstantiated. What we are comparing is not predictive success but predictive power, and N achieves predictive power comparable with E at the expense of predictively unsuccessful auxiliary hypotheses. The argument for the question-answering superiority of E confuses the ability of a theory to answer a question with its ability to answer correctly. Very likely E's answer is also incorrect.

It is to the informativeness of the failure of this reply that I have been leading. If the argument confuses the ability to answer with the ability to answer correctly, it is because that distinction is in fact confused. In a case such as

> (4) What is the amount of Mercury's precession?

which both E and N can answer with precision, the distinction is clear. But in a case such as (3), which asks for an explanation of an empirical finding, the distinction is unclear. What makes it unclear is that the epistemic credibility of a purported answer influences its status as an answer. N's purported answer to (3) consists in citing some auxiliary hypothesis, A, asserting, for example, the existence of heretofore undisclosed intra-Mercurial matter. If A is empirically falsified or discredited, it simply is not available to N for citation. The mere fact that A is logically consistent with the basic laws of N is insufficient. The ability of a theory to answer a question of type (3) amounts to its ability to explain a natural phenomenon or to solve an empirical problem. And the standards for explanation or solution preclude recourse to an independently discredited A. No theory is credited with achieving an explanation simply by virtue of complying with the logical canons of deduction. N attempted to answer (3) and some such attempts did produce deductions of the observed perturbations, but N was never credited with the ability to answer (3).

Questions of (3)'s type invariably require auxiliaries. The epistemic credentials of such auxiliaries can shift rapidly. Therefore, the question whether a theory can answer such a question is incomplete except as a retrospective question about a historical theory whose full career is available for review. Otherwise the question must specify a time, and the answer depends on the auxiliary information available at that time. Often the question of whether or not the presupposition of a question is compatible with a theory is only answered in the course of attempting to answer the question on the basis of that theory.

Grünbaum's discussion evades this historical dimension by encapsulating all auxiliary hypotheses involved in providing answers into the theory itself. But it is not a part of N, in the sense of a presupposition to which N is committed, that, to adapt an example of Grünbaum's, the only planets of nonnegligible gravitational influence on Uranus are Saturn and Jupiter. Otherwise N would be unable to answer the question of why Uranus exhibits certain perturbations, a question N did answer. In general, the question whether a theory can answer an empirical question remains unanswerable over significant periods in the theory's career. The important distinction is not between answers as such and correct answers, but between actual and potential answers. N had potential answers to (3), but none of them was realized.

For clarification of this point consider the following question:

(5) Why do Mercury's perturbations exceed the amount predicted by N?

Although (5) is so formulated that its presupposition explicitly conflicts with N, (5) is clearly legitimate for N. For (5) bears the same relation to N as

(6) Why do Uranus's perturbations exceed the amount predicted by N?

which N did answer. The difference between (5) and (6) with respect to N's question-answering capabilities is a difference in the confirmation of needed auxiliary hypotheses, a matter independent of N itself. N produced potential answers to both (5) and (6) but only the latter was realized.

Grünbaum generalizes his criticism of Popper by arguing that whenever the addition of an auxiliary hypothesis to a theory results in the correct postdiction of a phenomenon anomalous for the theory, no Popperian content increase can be said to occur. Let T_1 and T_2 be theories such that T_1 predicts a phenomenon e_1, a phenomenon e_2 incompatible with e_1 actually occurs, and T_2, which results from the addition of an auxiliary A_2 to T_1, postdicts e_2. Then the transition from T_1 to T_2 is not content-increasing by any Popperian nonmetrical comparison for the simple reason that T_1 and T_2 are mutually inconsistent. The inconsistency results from the occurrence in T_1 of a hypothesis A_1 inconsistent with A_2.

Now of course in its prediction of e_1, T_1 does invoke some such A_1. But this does not make A_1 an ingredient of T_1 such that any theory denying A_1 is a distinct theory inconsistent with T_1. A theory does not include, in the sense of part whose rejection implies rejection of the theory, any and all auxiliaries it uses in deducing empirical results. In fact, T_2 would not be considered a rival of T_1 at all, but a development of it. The original T_1 is not rejected in the transition to T_2, but salvaged by the introduction of A_2. A_1 was assumed implicitly by T_1, and this assumption was shown to be both incorrect and unnecessary for T_1. Thus, T_1 and T_2, insofar as they are distinguishable as theories at all, are certainly not mutually inconsistent. Grünbaum chides Popper for the naiveté of imagining the introduction of A_2 to be a "mere conjunctive appending" of A_2 to T_1, suggestive of increase in content. In fact, this operation alone would render T_2 inconsistent if A_1 were regarded as an ingredient of T_1. Undoubtedly, Popper did not so regard A_1, and in this he was surely justified. The status of A_2 in T_2 is different; A_2 does become an essential ingredient of T_2, given the empirical requirement that e_2 be postdicted. Thus, the contraction hypothesis did become an essential ingredient of the Lorentz electron theory, while its implicit denial in early versions of the theory was shown never to have been essential. Had it been, the theory could never have been reconciled with Michelson's result on pain of inconsistency. A content-increasing conjunctive appending seems to be exactly what occurred. Grünbaum's example of the introduction of Neptune is subject to a similar analysis, except that the Neptune hypothesis is more empirical and less theory-dependent than contraction, and so retains an auxiliary status.

Clarification of the nature of the operation by which auxiliaries are introduced to account for anomalies is of further importance in assessing Popper's use of truth and falsity content comparisons in his theory of verisimilitude. For that assessment has produced technical results whose significance is marred by the implausible rigidity of portraying theories as Tarskian deductive systems which set theoretically include all empirical predictions by which they are tested. For example, certain predictions of E do not, as is well known, accord exactly with observation. Corresponding predictions of N are in far greater disparity with observation. Thus, N's false predictions differ from E's false predictions. From this fact alone it immediately appears, without appeal to the usual technical apparatus, that the falsity content of E cannot be set-theoretically contained, properly or improperly, in that of N, violating one of Popper's conditions for attributing greater verisimilitude to E over N.[21] And, indeed, any incompatible theories yielding different predictions for a given observation will be incomparable as to (qualitative) Popperian verisimilitude unless one of them is confirmed. The difficulty with this apparently destructive result is that the appropriateness of including E's incorrect prediction in its falsity content class is a very complicated matter. For that prediction might well be corrected through changes of auxiliary information, leaving E itself unimpaired. The above reasoning precludes this possibility on pain of rendering E inconsistent.

Increasing explanatory and problem-solving power, as well as increasing predictive success, are properly regarded as forms of progress in science. Increase in relative truth of scientific hypotheses is the required explanans for all. This is not to say that relative truth is explicated by Popperian verisimilitude. If a theory itself is false, as opposed to yielding correctable, false predictions, then a trivial version of the above argument applies unambiguously to preclude its exceeding any other theory at least in qualitative Popperian verisimilitude. Content increase simply cannot be restricted to set-theoretic comparisons. Nor is it to say that all intuitive content increases exhibit these attributes. The Lorentz case is a counterexample, because increasing predictive success is a necessary condition for increases in the other attributes. A theory cannot take credit for an explanation or for a problem's solution, excepting problems that are purely pragmatic, unless the hypotheses in terms of which it purports to offer an explanation or solution are corroborated.

Scientists simply do not regard a problem as solved or a phenomenon as explained unless they believe that the epistemic situation entitles them to be confident that there is some truth to such hypotheses, that they are at least "partly right" or are "headed in the right direction." Insofar as the hypotheses are tentative, the task they address remains open. Insofar as conceptual problems are taken to require that they be treated instrumentally, only "systematization of experience" or solution of pragmatic problems, which amounts to predictive success alone, will be claimed as an achievement. The philosophical significance of this attitude is its reflection of the reluctance to accept the reality of phenomena that cannot be explained.

The further form of progress which I have claimed for science, its extension of the scope of observation, also requires the truthlikeness of theoretical hypotheses for its explanation. For theoretical entities are "directly observed" only if an approximately true theory precludes better evidence for their existence than that acquired in observing them. There is no sense to current talk of observations of elementary particles, for example, if no truth is imputed to those of their attributes that preclude less inferential access to them. The claim is not that there is no sense to talk of observing entities that are treated instrumentally; such talk could have an instrumental sense. The claim is that there would be no sense to the distinction drawn in science between direct and indirect observation if the entities said to be observed are treated instrumentally. But advance in the ability to observe directly, or advance from mere detection to observation, is an established form of progress in which the boundary of the observable shifts to encompass more phenomena. If there is no truth to theories postulating new entities, the progress apparent in the satisfaction of the criteria such theories fix for the observation of these entities is illusory.[22]

III

The case for increasing truth of scientific theories comes to this. There are these choices: (1) We can deny that science does exhibit what are normally taken to be its most manifest forms of progress; (2) we can deny the ultimate intelligibility of the world; (3) we can offer a realist explanation of progressive scientific change. (1) has the disadvantage of being at least as much in conflict with our experience as is realism with the historical record of theory change. (2) is always available. Skepticism as such remains a tenable epistemological position. But we must see the rejection of realism in this light. If we are prepared to deny that there is any truth to our most successful theories on the grounds that it is possible for a false theory to be successful, we might as well deny that there is an external world on the grounds that it is possible for a brain in a vat to have external-worldish experiences.

This is not to admit that the case for realism founders on skepticism. The argument I have given is of a form that may be parlayed into an attack on skepticism. It does not directly claim that predictive success of theoretical hypotheses attests to their truth. Rather, it claims that increases of predictive and explanatory success and of the scope of observation constitute explananda for a realist interpretation of science. Thus, the argument is indirect and hypothetical, as transcendental arguments must be. These features are double-edged. They enable the argument to meet the challenge to realism which the possibility of predictively successful, false theories poses. But they subject it to the inconclusiveness that plagues transcendental arguments generally.

The inclusiveness of the argument results partly from the continuing viability of (1) and (2), which some will find no more problematic than (3). (1) may be defended, for example, by arguing that extension of the scope of observation presents an explanans which itself is described in realist language, so that it is unsurprising that a realist explanans is forced upon us. It is possible to maintain that while the boundaries of the observable shift, theories retain "deep structures" which observation never penetrates, and which, accordingly, resist at least this motivation for a realist interpretation. It is possible to dismiss scientific talk of observation of theoretical entities altogether as a *façon de parler*, to reject the scientific distinction between experimental procedures that support hypotheses introducing theoretical entities and experimental procedures that constitute modes of observing such entities as a distinction without a difference. One explanandum for (3) then collapses.

I hope to have shown, however, that the others are free of the potential circularity of a realist explanation of phenomena described in realist terms. For if my argument against Grünbaum is right, realism is not required for *recognition* of advances in predictive, explanatory, or problem-solving power through scientific change. And if, as I have further argued, these explananda are persuasive on behalf of (3), it is at the very least an attractive fringe benefit of (3) that the scientific concept of observation is underwritten.

A further inconclusiveness results from the inability of a transcendental argument to establish its conclusion uniquely. I have sought to forestall this objection by offering an analysis of certain forms of scientific progress that requires the imputation of truth to hypotheses effecting such progress. Inasmuch as alternative analyses are possible,[23] it may be said that the impossibility of a nonrealist explanation of scientific progress has not been established.

There are, however, further advantages to (3) which any alternative short of (1) or (2) would be hard pressed to match. My argument for (3) may be applied, derivatively, at the level of individual theories. Realism with respect to individual theories is confirmed by its production of expectations beyond the explanatory function it is designed to serve, which in turn are satisfied. And this it does, for a successful theory characteristically provides explanations of phenomena and solutions of problems which emerge only as the theory is developed and applied in new areas, which are unanticipated in its initial formulation. How, short of (3), do we account for the ability of theories to explain and predict successfully phenomena outside the scope of the empirical laws they were designed to yield? This query answers Laudan's as to why we should denigrate ad hoc problem-solving devices.[24] The fact that 'ad hocness' is unfailingly pejorative is symptomatic of a realist attitude toward theories. Without realism there would be no rationalizing this attitude. We should welcome the introduction of new hypotheses, however contrived, which accommodate unanticipated, adverse evidence, rather than insist that solutions issue in a natural way from the original theory. Insofar as we need to add special hypotheses for emergent problems, a realist interpretation of the original theory is disconfirmed.

Still, it will be argued, it is not just the possibility of successful false theories with which realism must contend, but the fact of them. Is this any worse? Presumably so, for it shows that realism as a philosophical interpretation of science is not simply underdetermined but is refuted by the history of science. I wish to suggest, however, that history is not univocally opposed to realism any more than our experience of ordinary objects is univocally veridical. The difference in our epistemic stances with respect to scientific realism and other explanatory metaphysical doctrines is one of degree. For, apart from the moot issue of conceptual continuity through revolutions, one historical pattern that has remained stable throughout scientific change is the tenacity of preferential judgments about theories. Although a theory that replaces another is in turn replaced, its superiority over its predecessor continues to be recognized. As much as history records sustained judgments of the ultimate unacceptability of theories, it records sustained judgments of their relative merits.

Such judgments are not restricted to the pragmatic dimension of predictive success, but include explanatory comparisons. Newton provided a better explanation of free-fall than did Galileo although both explanations have been superseded. If we retain such judgments beyond the tenure of the theories themselves, we must regard one theory as having got more of the relevant facts right or as having described those facts more accurately, even if both theories are false. If the explanations proposed by both theories were rejected as utterly devoid of truth, such comparisons would be impossible. There seem to be as good inductive grounds for concluding that scientific theories increase in truth as for concluding that all theories are false.[25]

These conclusions are, moreover, consistent, if 'false' is taken to mean 'incompletely true'. Bivalence then requires distribution of truth values among logically independent components of theories. The above considerations regarding the status of auxiliary hypotheses and the technical difficulty of explicating truth approximation by set-theoretic comparison of truth and falsity contents of theories lead me to prefer a multivalued logic. In any case, multivalued logic is appropriate if the comparisons we wish to sustain pertain to degrees of accuracy. This seems a natural interpretation for statements of empirical science whose success in describing objective features of the world is not, intuitively, an all-or-nothing affair. It is, after all, quite common to compare descriptions, either verbal or pictorial, of a physical object as to accuracy without imagining that there is or could be such a thing as a maximally accurate description. Such a view does not require that there be an ultimate, absolute truth about any given aspect of the world, relative proximity to which explicates comparative truth assessments. In ethics and practical reasoning, we make comparative assessments of value without reference to an ideal state. It does not even require that truth be a relation between bearers of truth and the world, although this is the natural view if bearers of truth are taken to be statements. But it does require the independence of relative truth from the vicissitudes of justification and belief. For if the progressiveness of one theory over another is to be explained by its exceeding the other in truth value, then truth-value comparisons cannot be analyzed as epistemic comparisons.

Perhaps, however, the best way to block the skeptical induction is not to develop such a view but to resort to the philosophical ploy of self-reference which applies convincingly against verificationism and historical relativism and should, inductively, work again. For example: all philosophical theses of the past have turned out unacceptable (for who would claim as much progress for philosophy as for science at even its most theoretical level?). The claim that present theory is discredited by past theory is a philosophical thesis. Therefore . . .

NOTES

1. This is a strong version of the argument. If we suppose only that past theories have excelled by the standards of the time, we get a weaker version which invites recourse to a metalevel appraisal of the standards of different historical periods. Then we need metalevel standards and a meta-metalevel appraisal, and so forth; so we might as well stop with the formulation given.

2. I. Lakatos and A. Musgrave, eds., *Criticism and the Growth of Knowledge* (Cambridge: Cambridge University Press, 1970), 265.

3. Ibid., 189.

4. Lakatos distinguishes two senses of 'verisimilitude', Popper's technical sense and "intuitive truthlikeness," assimilating "classical verisimilitude" to the latter. The reader unwilling to believe that Lakatos could be guilty of so elementary a misunderstanding is encouraged to seek an alternative interpretation of ibid., 189. See also ibid., 265.

5. Cf. I. Hacking's review of Lakatos's *Collected Papers*, *British Journal for the Philosophy of Science*, vol. 30, no. 4.

6. *Criticism and the Growth of Knowledge*, 187.

7. L. Laudan, *Progress and its Problems* (Berkeley, Los Angeles, London: University of California Press, 1977).

8. Ibid., 12.

9. More precisely, Professor Laudan's view is that while neither truth nor falsity is determinable in principle, falsity but not truth can be estimated. A theory for which severe anomalies resist great effort at resolution is likely to be false, but no likelihood of truth can be imputed to a theory however impressive its achievements. Thus, Laudan is in a position to endorse a form of the skeptical, historical induction.

The claim underlying this view may be that the approximate falsity of a theory can be expected to issue in problem-solving failure, but approximate truth is no reason to expect problem-solving success. The grounds for such asymmetry, however, are unclear. It is insufficient to argue that extant philosophical analyses of approximate truth fail to support an inference to empirical success; neither do accounts of approximate falsity support an inference to empirical failure. Moreover, Laudan has shown in detail that false theories may be empirically successful (see his contribution to this volume).

On the other hand, the claim may be only the converse, that failure betokens falsehood. But according to Laudan, any degree of verisimilitude short of rigorous truth cannot be expected to issue in empirical success. Thus, a theory that fails may well be (very nearly) true.

10. D. Shapere, "The Character of Scientific Change," in *Scientific Discovery, Logic, and Rationality*, ed. T. Nickles, BSPS no. 56 (Dordrecht: D. Reidel, 1980), 61-102.

11. M. Levin, "On Theory-Change and Meaning-Change," *Philosophy of Science* 46, 3: 407-425.

12. There is, widespread among scientists at least, a notion of cumulativity very different from mine according to which theories retain their predecessors as limiting cases. This view, on every coherent version I know of, is refuted by the history of science.

13. See J. Leplin, "The Role of Models in Theory Construction," in *Scientific Discovery, Logic, and Rationality*, 267-283.

14. See Putnam's essay in this volume.

15. The "Principle of Charity" offered to secure the reference of past theories does nothing for their truth.

16. M. Hesse, "Truth and the Growth of Scientific Knowledge," *PSA* 1976, vol. 2, ed. F. Suppe and P. D. Asquith, pp. 261-281.

17. This is not to suppose as Putnam seems to, that just any form of empirical success requires realism for an explanation. There is no mystery, for example, about the ability of hypotheses proposed with the express purpose of providing a theoretical basis for antecedently established empirical laws to yield those laws, whether or not the hypotheses are interpreted realistically. For discussion of the forms of success requiring realist explanations, see J. Leplin, "The Historical Objection to Scientific Realism," *PSA* 1982, ed. T. Nichols and P. Asquith, pp. 88-98, and "Novel Prediction," in manuscript.

18. That this requirement is satisfied is argued in J. Leplin, "Reference and Scientific Realism," *Studies in History and Philosophy of Science*, 10, 4: 265-285, which however, presupposes in opposition to positivism that evidence for a theory is evidence for the theory as a whole, not just for its observational portions.

19. A. Grünbaum, "Can A Theory Answer More Questions than One of Its Rivals?" *British Journal for the Philosophy of Science*, 27: 1-23.

20. Professor Laudan points out that in the general case such an assumption will be very problematic. Changes of ontology, for example, will shift radically the ranges of questions to be addressed. And a shift that reduces the number of questions answered may well be progressive if achieved through ontological economy.

21. This elementary observation is amply illustrated by using Popper's familiar example of comparing false estimates of clock-time. Let the clock read 9:48 and let the interval estimates exclude upper bounds. Then the estimate "it is between 9:45 and 9:48" fails to exceed the estimate "it is between 9:40 and 9:48" in Popperian verisimilitude since, for example, it implies the falsehood "it is between 9:44 and 9:48" which is not implied by the latter.

22. For further argument, see J. Leplin, "The Assessment of Auxiliary Hypotheses," *British Journal for the Philosophy of Science* 33 (1982): 235-249, and "Methodological Realism and Scientific Rationality," forthcoming in *Philosophy of Science*.

The truthlikeness of theoretical hypotheses, has, arguably, further explananda which I neglect as more controversial and less obviously forms of progress, for example, conceptual continuity through revolutionary changes in mature science.

23. It is noteworthy that such a nonrealist as Laudan, who does recognize explanatory or problem-solving advances through theory change, offers no analysis of what constitutes a solution to a scientific problem beyond requiring a deductive relationship between a theory and a statement of the problem. Rather, he defers to what scientists regard as solutions. Cf. *Progress and its Problems*, 22.

24. Ibid., 115.

25. There are cases of theories that continue to be regarded as explanatory without apparent continuing imputation of truth to the explanans. In such cases, the explanandum also is rejected. The problem is of the theory's own making (or is the creation of a predecessor) and subsequent theory dissolves rather than solves it. Explanatory comparisons among such cases are to be excluded from the indicated induction.

11

A Confutation of Convergent Realism

Larry Laudan

> The positive argument for realism is that it is the only philos-
> ophy that doesn't make the success of science a miracle.
> —H. Putnam (1975)

THE PROBLEM

It is becoming increasingly common to suggest that epistemological realism is an empirical hypothesis, grounded in, and to be authenticated by,
its ability to explain the workings of science. A growing number of philosophers (including Boyd, W. Newton-Smith, A. Shimony, Putnam,
and I. Niiniluoto) have argued that the theses of epistemic realism are
open to empirical test.[1] The suggestion that epistemological doctrines
have much the same empirical status as the sciences is a welcome one;
for, whether it stands up to detailed scrutiny or not, this suggestion
marks a significant facing-up by the philosophical community to one of
the most neglected (and most notorious) problems of philosophy: the
status of epistemological claims.

There are, however, potential hazards as well as advantages associated
with the "scientizing" of epistemology. Specifically, once one concedes
that epistemic doctrines are to be tested in the court of experience, it is
possible that one's favorite epistemic theories may be refuted rather than
confirmed. It is the thesis of this paper that precisely such a fate afflicts a

218

form of realism advocated by those who have been in the vanguard of the move to show that realism is supported by an empirical study of the development of science. Specifically, I will show that epistemic realism, at least in certain of its extant forms, is neither supported by, nor has it made sense of, much of the available historical evidence.

CONVERGENT REALISM

Like other philosophical "isms," the term 'realism' covers a variety of sins. Many of these will not be at issue here. For instance, 'semantic realism' (in brief, the claim that all theories have truth values and that some theories are true, although we know not which) is not in dispute. Nor shall I discuss what one might call 'intentional realism' (i.e., the view that theories are generally intended by their proponents to assert the existence of entities corresponding to the terms in those theories). What I will focus on instead are certain forms of epistemological realism. As Hilary Putnam has pointed out, although such realism has become increasingly fashionable, "very little is said about what realism *is*."[2] The lack of specificity about what realism asserts makes it difficult to evaluate its claims, since many formulations are too vague and sketchy to get a grip on. At the same time, any efforts to formulate the realist position with greater precision lay the critic open to charges of attacking a straw man. In the course of this paper, I shall attribute several theses to the realists. Although there is probably no realist who subscribes to all of them, most of them have been defended by some self-avowed realist or other; taken together, they are perhaps closest to that version of realism advocated by Putnam, Boyd, and Newton-Smith. Although I believe the views I shall be discussing can be legitimately attributed to certain contemporary philosophers (and I will cite the textual evidence for such attributions), it is not crucial to my case that such attributions can be made. Nor will I claim to do justice to the complex epistemologies of those whose work I will criticize. Rather, my aim is to explore certain epistemic claims which those who are realists might be tempted (and in some cases have been tempted) to embrace. If my arguments are sound, we will discover that some of the most intuitively tempting versions of realism prove to be chimeras.

The form of realism I shall discuss involves variants of the following claims:

(R1) Scientific theories (at least in the 'mature' sciences) are typically approximately true, and more recent theories are closer to the truth than older theories in the same domain.

(R2) The observational and theoretical terms within the theories of a mature science genuinely refer (roughly, there are substances in the world that correspond to the ontologies presumed by our best theories).

(R3) Successive theories in any mature science will be such that they preserve the theoretical relations and the apparent referents of earlier theories, that is, earlier theories will be limiting cases of later theories.[3]

(R4) Acceptable new theories do and should explain why their predecessors were successful insofar as they were successful.

To these semantic, methodological, and epistemic theses is conjoined an important metaphilosophical claim about how realism is to be evaluated and assessed. Specifically, it is maintained that:

(R5) Theses (R1) to (R4) entail that ("mature") scientific theories should be successful; indeed, these theses constitute the best, if not the only, explanation for the success of science. The empirical success of science (in the sense of giving detailed explanations and accurate predictions) accordingly provides striking empirical confirmation for realism.

I shall call the position delineated by (R1) to (R5) *convergent epistemological realism*, or CER for short. Many recent proponents of CER maintain that (R1), (R2), (R3), and (R4) are empirical hypotheses that, via the linkages postulated in (R5), can be tested by an investigation of science itself. They propose two elaborate abductive arguments. The structure of the first (argument 1) which is germane to (R1), is something like this:

1. If scientific theories are approximately true, then they typically will be empirically successful.
2. If the central terms in scientific theories genuinely refer, then those theories generally will be empirically successful.
3. Scientific theories are empirically successful.
4. (Probably) theories are approximately true and their terms genuinely refer.

The structure of the second abductive argument (argument 2), which is relevant to (R3), is of slightly different form, specifically:

1. If the earlier theories in a "mature" science are approximately true, and if the central terms of those theories genuinely refer, then later, more successful theories in the same science will preserve the earlier theories as limiting cases.
2. Scientists seek to preserve earlier theories as limiting cases and generally succeed in doing so.
3. (Probably) earlier theories in a "mature" science are approximately true and genuinely referential.

Taking the success of present and past theories as givens, proponents of CER claim that *if* CER were true, it would follow, as a matter of course, that science would be successful and progressive. Equally, they allege that if CER were false, the success of science would be "miraculous" and without explanation.[4] Because (on their view) CER explains the fact that science is successful, the theses of CER are thereby confirmed by the success of science, and nonrealist epistemologies are discredited by the latter's alleged inability to explain both the success of current theories and the progress which science historically exhibits.

As Putnam and certain others (e.g., Newton-Smith) see it, the fact that statements about reference (R2, R3) or about approximate truth (R1, R3) function in the explanation of a contingent state of affairs, establishes that "the notions of 'truth' and 'reference' have a causal explanatory role in epistemology."[5] In one fell swoop, both epistemology and semantics are 'naturalized' and, to top it all off, we get an explanation of the success of science thrown into the bargain!

The central question before us is whether the realist's assertions about the interrelations between truth, reference, and success are sound. It will be the burden of this paper to raise doubts about both arguments 1 and 2. Specifically, I will argue that four of the five premises of those abductions are either false or too ambiguous to be acceptable. I will also seek to show that, even if the premises were true, they would not warrant the conclusions that realists draw from them. The next three sections of this essay deal with the first abductive argument. Then I turn to the second.

REFERENCE AND SUCCESS

The specifically referential side of the empirical argument for realism has been developed chiefly by Putnam, who talks explicitly of reference rather more than most realists. However, reference is usually implicitly smuggled in, since most realists subscribe to the (ultimately referential) thesis that "the world probably contains entities very like those postulated by our most successful theories."

If (R2) is to fulfill Putnam's ambition that reference can explain the success of science, and that the success of science establishes the presumptive truth of (R2), it seems he must subscribe to claims similar to these:

(S1) The theories in the advanced or mature sciences are successful.

(S2) A theory whose central terms genuinely refer will be a successful theory.

(S3) If a theory is successful, we can reasonably infer that its central terms genuinely refer.

(S4) All the central terms in theories in the mature sciences do refer.

There are complex interconnections here. (S2) and (S4) explain (S1), while (S1) and (S3) provide the warrant for (S4). Reference explains success, and success warrants a presumption of reference. The arguments are plausible, given the premises. But there is the rub, for with the possible exception of (S1), none of the premises is acceptable.

The first and toughest problem involves getting clearer about the nature of that "success" which realists are concerned to explain. Although Putnam, W. Sellars, and Boyd all take the success of certain sciences as a given, they say little about what this success amounts to. So far as I can see, they are working with a largely *pragmatic* notion to be couched in terms of a theory's workability or applicability. On this account, we would say that a theory is successful if it makes substantially correct predictions, if it leads to efficacious interventions in the natural order, and if it passes a battery of standard tests. One would like to be able to be more specific about what success amounts to, but the lack of a coherent theory of confirmation makes further specificity very difficult.

Moreover, the realist must be wary, at least for these purposes, of adopting too strict a notion of success, for a highly robust and stringent construal of "success" would defeat the realist's purposes. What he wants to explain, after all, is why science in general has worked so well. If he were to adopt a very demanding characterization of success (such as those advocated by inductive logicians or Popperians), then it would probably turn out that science has been largely "unsuccessful" (because it does not have high confirmation), and the realist's avowed explanandum would thus be a nonproblem. Accordingly, I will assume that a theory is successful so long as it has worked well, that is, so long as it has functioned in a variety of explanatory contexts, has led to confirmed predictions, and has been of broad explanatory scope. As I understand the realist's position, his concern is to explain why certain theories have enjoyed this kind of success.

If we construe 'success' in this way, (S1) can be conceded. Whether one's criterion of success is broad explanatory scope, possession of a large number of confirming instances, or conferring manipulative or predictive control, it is clear that science, by and large, is a successful activity.

What about (S2)? I am not certain that any realist would or should endorse it, although it is a perfectly natural construal of the realist's claim that "reference explains success." The notion of reference that is involved here is highly complex and unsatisfactory in significant respects. Without endorsing it, I shall use it frequently in the ensuing discussion. The realist sense of reference is a rather liberal one, according to which the terms in a theory may be genuinely referring even if many of the claims the theory makes about the entities to which it refers are false. Provided that there are entities that "approximately fit" a theory's description of them, Putnam's charitable account of reference allows us to say that the terms of a theory genuinely refer.[6] On this account (and these are Putnam's examples), Bohr's 'electron', Newton's 'mass', Mendel's 'gene', and Dalton's 'atom' are all referring terms, while 'phlogiston' and 'ether' are not.[7]

Are genuinely referential theories (i.e., theories whose central terms genuinely refer) invariably or even generally successful at the empirical level, as (S2) states? There is ample evidence that they are not. The chemical atomic theory in the eighteenth century was so remarkably unsuccessful that most chemists abandoned it in favor of a more phenomenological, elective affinity chemistry. The Proutian theory that the atoms of heavy elements are composed of hydrogen atoms had, through most of the nineteenth century, a strikingly unsuccessful career, confronted by a long string of apparent refutations. The Wegenerian theory that the continents are carried by large subterranean objects moving laterally across the earth's surface was, for some thirty years in the recent history of geology, a strikingly unsuccessful theory until, after major modifications, it became the geological orthodoxy of the 1960s and 1970s. Yet all of these theories postulated basic entities which (according to Putnam's "principle of charity") genuinely exist.

The realist's claim that we should expect referring theories to be empirically successful is simply false. And, with a little reflection, we can see good reasons why it should be. To have a genuinely referring theory is to have a theory that "cuts the world at its joints," a theory that postulates entities of a kind that really exist. But a genuinely referring theory need not be such that all—or even most—of the specific claims it makes about the properties of those entities and their modes of interaction are true. Thus, Dalton's theory makes many false claims about atoms; Bohr's early theory of the electron was similarly flawed in important respects. Contra-(S2), genuinely referential theories need not be strikingly successful, since such theories may be 'massively false' (i.e., have far greater falsity content than truth content).

(S2) is so patently false that it is difficult to imagine that the realist need be committed to it. But what else will do? The (Putnamian) realist wants attributions of reference to a theory's terms to function in an explanation of that theory's success. The simplest and crudest way of doing that involves a claim like (S2). A less outrageous way of achieving the same end would involve the weaker:

(S2') A theory whose terms refer will usually (but not always) be successful.

Isolated instances of referring but unsuccessful theories, sufficient to refute (S2), leave (S2') unscathed. But, if we were to find a broad range of referring but unsuccessful theories, that would be evidence against (S2′). Such theories can be generated at will. For instance, take any set of terms which one believes to be genuinely referring. In any language rich enough to contain negation, it will be possible to construct indefinitely many unsuccessful theories, all of whose substantive terms are genuinely referring. Now, it is always open to the realist to claim that such "theories" are not really theories at all, but mere conjunctions of isolated statements—lacking that sort of conceptual integration we associate with "real" theories. Sadly, a parallel argument can be made for genuine theories. Consider, for instance, how many inadequate versions of the atomic theory there were in the 2,000 years of atomic speculating, before a genuinely successful theory emerged. Consider how many unsuccessful versions there were of the wave theory of light before the 1820s, when a successful wave theory first emerged. Kinetic theories of heat in the seventeenth and eighteenth century, and developmental theories of embryology before the late nineteenth century, sustain a similar story. (S2′), every bit as much as (S2), seems hard to reconcile with the historical record.

As Richard Burian has pointed out to me (personal communication), a realist might attempt to dispense with both of those theses and simply rest content with (S3) alone. Unlike (S2) and (S2′), (S3) is not open to the objection that referring theories are often unsuccessful, for it makes no claim that referring theories are always or generally successful. But (S3) has difficulties of its own. In the first place, it seems hard to square with the fact that the central terms of many relatively successful theories (e.g., ether theories or phlogistic theories) are evidently nonreferring. I shall discuss this tension in detail below. More crucial for our purposes here is that (S3) is *not strong enough* to permit the realist to utilize reference to explain success. Unless genuineness of reference entails that all or most referring theories will be successful, then the fact that a theory's terms refer scarcely provides a convincing explanation of that theory's success. If, as (S3) allows, many (or even most) referring theories can be unsuccessful, how can the fact that a successful theory's terms refer be taken to explain why it is successful? (S3) may or may not be true; but in either case it arguably gives the realist no explanatory access to scientific success.

A more plausible construal of Putnam's claim that reference plays a role in explaining the success of science involves a rather more indirect argument. It might be said (and Putnam does say this much) that we can explain why a theory is successful by assuming that the theory is true or approximately true. Since a theory can only be true or nearly true (in any sense of those terms open to the realist) if its terms genuinely refer, it might be argued that reference gets into the act willy-nilly when we explain a theory's success in terms of its truthlike status. On this account, reference is piggybacked on approximate truth. The viability of this indirect approach is treated at length in the next section, so I will not discuss it here except to observe that if the only contact point between reference and success is provided through the medium of approximate truth, then the link between reference and success is extremely tenuous.

What about (S3), the realist's claim that success creates a rational presumption of reference? We have already seen that (S3) provides no explanation of the success of science, but does it have independent merits? The question specifically is whether the success of a theory provides a warrant for concluding that its central terms refer. Insofar as this is, as certain realists suggest, an empirical question, it requires us to inquire whether past theories which have been successful are ones whose central terms genuinely referred (according to the realist's own account of reference).

A proper empirical test of this hypothesis would require an extensive sifting of the historical record that is not possible to perform here. What I can do is to mention a range of once successful, but (by present lights) nonreferring, theories. A fuller list will come later, but for now we will focus on a whole family of related theories, namely, the subtle fluids and ethers of eighteenth- and nineteenth-century physics and chemistry.

Consider specifically the state of etherial theories in the 1830s and 1840s. The electrical fluid, a substance that was generally assumed to accumulate on the surface rather than to permeate the interstices of bodies, had been utilized to explain inter alia the attraction of oppositely charged bodies, the behavior of the Leyden jar, the similarities between atmospheric and static electricity, and many phenomena of current electricity. Within chemistry and heat theory, the caloric ether had been widely utilized since H. Boerhaave (by, among others, A. L. Lavoisier, P. S. Laplace, J. Black, Count Rumford, J. Hutton, and H. Cavendish) to explain everything from the role of heat in chemical reactions to the conduction and radiation of heat and several standard problems of thermometry. Within the theory of light, the optical ether functioned centrally in explanations of reflection, refraction, interference, double refraction, diffraction, and polarization. (Of more than passing interest, optical ether theories had also made some very startling predictions, e.g., A. Fresnel's prediction of a bright spot at the center of the shadow of a circular disc; a surprising prediction which, when tested, proved correct. If that does not count as empirical success, nothing does!) There were also gravitational (e.g., G. LeSage's) and physiological (e.g., D. Hartley's) ethers which enjoyed some measure of empirical success. It would be difficult to find a family of theories in this period as successful as ether theories; compared with them, nineteenth-century atomism (for instance), a genuinely referring theory (on realist accounts), was a dismal failure. Indeed, on any account of empirical success which I can conceive of, nonreferring nineteenth-century ether theories were more successful than contemporary, referring atomic theories. In this connection, it is worth recalling the remark of the great theoretical physicist, J. C. Maxwell, to the effect that the ether was better confirmed than any other theoretical entity in natural philosophy.

What we are confronted with in nineteenth-century ether theories, then, is a wide variety of once successful theories, whose central explanatory concept Putnam singles out as a prime example of a nonreferring one.[8] What are (referential) realists to make of this historical case? On the face of it, this case poses two rather different kinds of challenges to realism: first, it suggests that (S3) is a dubious piece of advice in that *there can be* (and have been) *highly successful theories some central terms of which are nonreferring;* and second, it suggests that *the realist's claim that he can explain why science is successful is false at least insofar as a part of the historical success of science has been success exhibited by theories whose central terms did not refer.*

But perhaps I am being less than fair when I suggest that the realist is committed to the claim that *all* the central terms in a successful theory refer. It is possible than when Putnam, for instance, says that "terms in a mature [or successful] science typically refer,"[9] he only means to suggest that *some* terms in a successful theory or science genuinely refer. Such a claim is fully consistent with the fact that certain other terms (e.g., 'ether') in certain successful, mature sciences (e.g., nineteenth-century physics) are nonetheless nonreferring. Put differently, the realist might argue that the success of a theory warrants the claim that at least some (but not necessarily all) of its central concepts refer.

Unfortunately, such a weakening of (S3) entails a theory of evidential support which can scarcely give comfort to the realist. After all, part of what separates the realist from the positivist is the former's belief that the evidence for a theory is evidence for *everything* the theory asserts. Where the stereotypical positivist argues that the evidence selectively confirms only the more 'observable' parts of a theory, the realist generally asserts (in the language of Boyd) that:

the sort of evidence which ordinarily counts in favor of the acceptance of a scientific law or theory is, ordinarily, evidence for the (at least approximate) truth of the law or theory as an account of the causal relations obtaining between the entities ["observational or theoretical"] quantified over in the law or theory in question.[10]

For realists such as Boyd, either all parts of a theory (both observational and nonobservational) are confirmed by successful tests or none are. In general, realists have been able to utilize various holistic arguments to insist that it is not merely the lower-level claims of a well-tested theory that are confirmed but its deep-structural assumptions as well. This tactic has been used to good effect by realists in establishing that inductive support 'flows upward' so as to authenticate the most 'theoretical' parts of our theories. Certain latter-day realists (e.g., Glymour) want to break out of this holist web and argue that certain components of theories can be 'directly' tested. This approach runs the very grave risk of undercutting what the realist desires most: a rationale for taking our deepest-structure theories seriously, and a justification for linking reference and success. After all, if the tests to which we subject our theories only test *portions* of those theories, then even highly successful theories may well have central terms that are nonreferring and central tenets that, because untested, we have no grounds for believing to be approximately true. Under those circumstances, a theory might be highly successful and yet contain important constituents that were patently false. Such a state of affairs would wreak havoc with the realist's presumption (thesis R1) that success betokens approximate truth. In short, to be less than a holist about theory testing is to put at risk precisely that predilection for deep-structure claims which motivates much of the realist enterprise.

There is, however, a rather more serious obstacle to this weakening of referential realism. It is true that by weakening (S3) to only certain terms in a theory, one would immunize it from certain obvious counterexamples. But such a maneuver has debilitating consequences for other central realist theses. Consider the realist's thesis (R3) about the retentive character of intertheory relations (discussed below in detail). The realist both recommends as a matter of policy and claims as a matter of fact that successful theories are (and should be) rationally replaced only by theories that preserve reference for the central terms of their successful predecessors. The rationale for the normative version of this retentionist doctrine is that the terms in the earlier theory, *because it was successful, must have been referential,* and thus a constraint on any successor to that theory is that reference should be retained for such terms. This makes sense just in case success provides a blanket warrant for presumption of reference. But if (S3) were weakened so as to say merely that it is reasonable to assume that *some* of the terms in a successful theory genuinely refer, then the realist would have no rationale for his retentive theses (variants of R3), which have been a central pillar of realism for several decades.[11]

Something apparently has to give. A version of (S3) strong enough to license (R3) seems incompatible with the fact that many successful theories contain nonreferring central terms. But any weakening of (S3) dilutes the force of, and removes the rationale for, the realist's claims about convergence, retention, and correspondence in intertheory relations.[12] If the realist once concedes that some unspecified set of the terms of a successful theory may well not refer, then his proposals for restricting "the class of candidate theories" to those that retain reference for the prima facie referring terms in earlier theories is without foundation.[13]

More generally, we seem forced to say that such linkages as there are between reference and success are rather murkier than Putnam's and Boyd's discussions would lead us to believe. If the realist is going to make his case for CER, it seems that it will have to hinge on approximate truth, (R1), rather than reference, (R2).

APPROXIMATE TRUTH AND SUCCESS: THE DOWNWARD PATH

Ignoring the referential turn among certain recent realists, most realists continue to argue that, at bottom, epistemic realism is committed to the view that successful scientific theories, even if strictly false, are nonetheless 'approximately true' or 'close to the truth' or 'verisimilar'.[14] The claim generally amounts to this pair:

(T1) If a theory is approximately true, then it will be explanatorily successful.

(T2) If a theory is explanatorily successful, then it is probably approximately true.

What the realist would *like* to be able to say, of course, is:

(T1') If a theory is true, then it will be successful.

(T1') is attractive because self-evident. Most realists, however, balk at invoking (T1') because they are (rightly) reluctant to believe that we can reasonably presume of any given scientific theory that it is true. If all the realist could explain was the success of theories that were true *simpliciter*, his explanatory repertoire would be acutely limited. As an attractive move in the direction of broader explanatory scope, (T1) is rather more appealing. After all, presumably many theories which we believe to be false (e.g., Newtonian mechanics, thermodynamics, wave optics) were— and still are—highly successful across a broad range of applications.

Perhaps, the realist evidently conjectures, we can find an *epistemic* account of that pragmatic success by assuming such theories to be 'approximately true'. But we must be wary of this potential sleight of hand. It may be that there is a connection between success and approximate truth; *but if there is such a connection it must be independently argued for.* The acknowledgedly uncontroversial character of (T1') must not be surreptitiously invoked—as it sometimes seems to be—in order to establish T1. When the antecedent of (T1') is appropriately weakened by speaking of approximate truth, it is by no means clear that (T1) is sound.

Virtually all the proponents of epistemic realism take it as unproblematic that if a theory were approximately true, it would deductively follow that the theory would be a relatively successful predictor and explainer of observable phenomena. Unfortunately, few of the writers of whom I am aware have defined what it means for a statement or theory to be 'approximately true'. Accordingly, it is impossible to say whether the alleged entailment is genuine. This reservation is more than perfunctory. Indeed, on the best known account of what it means for a theory to be approximately true, it does *not* follow that an approximately true theory will be explanatorily successful.

Suppose, for instance, that we were to say in a Popperian vein that a theory, T_1, is approximately true if its truth content is greater than its falsity content, that is,

$$Ct_T(T_1) \gg Ct_F(T_1),^{15}$$

where $Ct_T(T_1)$ is the cardinality of the set of true sentences entailed by T_1, and $Ct_F(T_1)$ is the cardinality of the set of false sentences entailed by T_1. When approximate truth is so construed, it does *not* logically follow that an arbitrarily selected class of a theory's entailments (namely, some of its observable consequences) will be true. Indeed, it is entirely conceivable that a theory might be approximately true in the indicated sense and yet be such that *all* of its consequences tested thus far are *false*.[16]

Some realists concede their failure to articulate a coherent notion of approximate truth or verisimilitude but insist that this failure in no way compromises the viability of (T1). Newton-Smith, for instance, grants that "no one has given a satisfactory analysis of the notion of verisimilitude,"[17] but insists that the concept can be legitimately invoked "even if one cannot at the time give a philosophically satisfactory analysis of it." He quite rightly points out that many scientific concepts were explanatorily useful long before a philosophically coherent analysis was given for them. But the analogy is unseemly, for what is being challenged is not whether the concept of approximate truth is philosophically rigorous, but rather whether it is even clear enough for us to ascertain whether it entails what it purportedly explains. Until someone provides a clearer analysis of approximate truth than is now available, it is not even clear whether truthlikeness would explain success, let alone whether, as

Newton-Smith insists,[18] "the concept of verisimilitude is *required* in order to give a satisfactory theoretical explanation of an aspect of the scientific enterprise." If the realist would demystify the "miraculousness" (Putnam) or the "mysteriousness" (Newton-Smith) of the success of science, he needs more than a promissory note that somehow, someday, someone will show that approximately true theories must be successful theories.[19]

It is not clear whether there is some definition of approximate truth that does indeed entail that approximately true theories will be predictively successful (and yet still probably false).[20] What can be said is that, promises to the contrary notwithstanding, none of the proponents of realism has yet articulated a coherent account of approximate truth which entails that approximately true theories will, across the range where we can test them, be successful predictors. Further difficulties abound. Even if the realist had a semantically adequate characterization of approximate or partial truth, and even if that semantics entailed that most of the consequences of an approximately true theory would be true, he would still be without any criterion that would *epistemically* warrant the ascription of approximate truth to a theory. As it is, the realist seems to be long on intuitions and short on either a semantics or an epistemology of approximate truth.

These should be urgent items on the realists' agenda since, until we have a coherent account of what approximate truth is, central realist theses such as (R1), (T1), and (T2) are just so much mumbo jumbo.

APPROXIMATE TRUTH AND SUCCESS: THE UPWARD PATH

Despite the doubts voiced in the previous section, let us grant for the sake of argument that if a theory is approximately true, then it will be successful. Even granting (T1), is there any plausibility to the suggestion of (T2) that explanatory success can be taken as a rational warrant for a judgment of approximate truth? The answer seems to be "no."

To see why, we need to explore briefly one of the connections between "genuinely referring" and being "approximately true." However the latter is understood, I take it that *a realist would never want to say that a theory was approximately true if its central terms failed to refer.* If there were nothing like genes, then a genetic theory, no matter how well confirmed it was, would not be approximately true. If there were no entities similar to atoms, no atomic theory could be approximately true; if there were no subatomic particles, then no quantum theory of chemistry could be approximately true. In short, a necessary condition, especially for a scientific realist, for a theory being close to the truth is that its central explanatory terms genuinely refer. (An *instrumentalist*, of course, could countenance the weaker claim that a theory was approximately true so long as its directly testable consequences were close to the observable values. But as I argued above, the realist must take claims about approximate truth to refer alike to the observable and the deep-structural dimensions of a theory.)

Now, what the history of science offers us is a plethora of theories that were both successful and (so far as we can judge) nonreferential with respect to many of their central explanatory concepts. I discussed earlier one specific family of theories that fits this description. Let me add a few more prominent examples to the list:

—the crystalline spheres of ancient and medieval astronomy;
—the humoral theory of medicine;
—the effluvial theory of static electricity;
—"catastrophist" geology, with its commitment to a universal (Noachian) deluge;
—the phlogiston theory of chemistry;
—the caloric theory of heat;
—the vibratory theory of heat;
—the vital force theories of physiology;
—the electromagnetic ether;
—the optical ether;
—the theory of circular inertia; and
—theories of spontaneous generation.

This list, which could be extended ad nauseam, involves in every case a theory that was once successful and well confirmed, but which contained central terms that (we now believe) were nonreferring. Anyone who imagines that the theories that have been successful in the history of science have also been, with respect to their central concepts, genuinely referring theories has studied only the more whiggish versions of the history of science (i.e., the ones which recount only those past theories that are referentially similar to currently prevailing ones).

It is true that proponents of CER sometimes hedge their bets by suggesting that their analysis applies exclusively to "the mature sciences" (e.g., Putnam and W. Krajewski). This distinction between mature and immature sciences proves convenient to the realist since he can use it to dismiss any prima facie counterexample to the empirical claims of CER on the grounds that the example is drawn from a so-called immature science. But this insulating maneuver is unsatisfactory in two respects. In the first place, it runs the risk of making CER vacuous since these authors generally define a mature science as one in which correspondence or limiting-case relations obtain invariably between any successive theories in the science once it has passed "the threshold of maturity." Krajewski grants the tautological character of this view when he notes that "the thesis that there is [correspondence] among successive theories becomes, indeed, analytical."[21] Nonetheless, he believes that there is a version of the maturity thesis which "may be and must be tested by the history of science." That version is that "every branch of science crosses at some period the threshold of maturity." But the testability of this hypothesis is dubious at best. There is no historical observation that could conceivably *refute* it since, even if we discovered that no sciences yet possessed "corresponding" theories, it could be maintained that eventually every science will become corresponding. It is equally difficult to *confirm* it since, even if we found a science in which corresponding relations existed between the latest theory and its predecessor, we would have no way of knowing whether that relation will continue to apply to subsequent changes of theory in that science. In other words, the much-vaunted empirical testability of realism is seriously compromised by limiting it to the mature sciences.

But there is a second unsavory dimension to the restriction of CER to the mature sciences. The realists' avowed aim, after all, is to explain why science is successful: that is the "miracle" they allege the nonrealists leave unaccounted for. The fact of the matter is that parts of science, including many immature sciences, have been successful for a very long time; indeed, many of the theories I alluded to above were empirically successful by any criterion I can conceive of (including fertility, intuitively high confirmation, successful prediction, etc.). If the realist restricts himself to explaining only how the mature sciences work (and recall that very few sciences indeed are yet mature as the realist sees it), then he will have completely failed in his ambition to explain why science in general is successful. Moreover, several of the examples I have cited above come from the history of mathematical physics in the last century (e.g., electromagnetic and optical ethers) and, as Putnam himself concedes, "*physics* surely counts as a 'mature' science if any science does."[22] Since realists would presumably insist that many of the central terms of the theories enumerated above do not genuinely refer, it follows that none of those theories could be approximately true (recalling that the former is a necessary condition for the latter). Accordingly, cases of this kind cast very grave doubts on the plausibility of (T2), that is, the claim that nothing succeeds like approximate truth.

I daresay that for every highly successful theory in the history of science that we now believe to be a genuinely referring theory, one could find half a dozen once successful theories that we now regard as substantially nonreferring. If the proponents of CER are the empiricists they profess to be about matters epistemological, cases of this kind and this frequency should give them pause about the well-foundedness of (T2).

But we need not limit our counterexamples to nonreferring theories. There were many theories in the past that (so far as we can tell) were both genuinely referring and empirically successful which we are nonetheless loathe to regard as approximately true. Consider, for instance, virtually all those geological theories prior to the 1960s which denied any lateral motion to the continents. Such theories were, by any standard, highly successful (and apparently referential); but would anyone today be prepared to say that their constituent theoretical claims—committed as they were to laterally stable continents—are almost true? Is it not the fact of the matter that structural geology was a successful science between (say) 1920 and 1960, even though geologists were fundamentally mistaken about many (perhaps even most) of the basic mechanisms of tectonic construction? Or what about the chemical theories of the 1920s which assumed that the atomic nucleus was structurally homogenous? Or those chemical and physical theories of the late nineteenth century which explicitly assumed that matter was neither created nor destroyed? I am aware of no sense of approximate truth (available to the realist) according to which such highly successful, but evidently false, theoretical assumptions could be regarded as "truthlike."

More generally, the realist needs a riposte to the prima facie plausible claim that there is no necessary connection between increasing the accuracy of our deep-structural characterizations of nature and improvements at the level of phenomenological explanations, predictions, and manipulations. It *seems* entirely conceivable intuitively that the theoretical mechanisms of a new theory, T_2, might be closer to the mark than those of a rival T_1, and yet T_1 might be more accurate at the level of testable predictions. In the absence of an argument that greater correspondence at the level of unobservable claims is more likely than not to reveal itself in greater accuracy at the experimental level, one is obliged to say that the realist's hunch that increasing deep-structural fidelity must manifest itself pragmatically in the form of heightened experimental accuracy has yet to be made cogent. (Equally problematic, of course, is the inverse argument to the effect that increasing experimental accuracy betokens greater truthlikeness at the level of theoretical, i.e., deep-structural, commitments.)

CONFUSIONS ABOUT CONVERGENCE AND RETENTION

Thus far, I have discussed only the static or synchronic versions of CER, versions that make absolute rather than relative judgments about truthlikeness. Of equal appeal have been those variants of CER that invoke a notion of what is variously called "convergence," "correspondence," or "cumulation." Proponents of the diachronic version of CER supplement the arguments discussed above [(S1)-(S4) and (T1)-(T2)] with an additional set. They tend to be of this form:

(C1) If earlier theories in a scientific domain are successful and thereby, according to realist principles [e.g., (S3) above], approximately true, then scientists should only accept later theories that retain appropriate portions of earlier theories.

(C2) As a matter of fact, scientists do adopt the strategy of (C1) and manage to produce new, more successful theories in the process.

(C3) The "fact" that scientists succeed at retaining appropriate parts of earlier theories in more successful successors shows that the earlier theories did genuinely refer and that they are approximately true. And thus, the strategy propounded in (C1) is sound.[23]

Perhaps the prevailing view here is Putnam's and (implicitly) Popper's, according to which rationally warranted successor theories in a mature science must contain reference to the entities apparently referred to in the predecessor theory (since, by hypothesis, the terms in the earlier theory refer), and also contain the theoretical laws and mechanisms of the predecessor theory as limiting cases. As Putnam tells us, a realist should insist that *any* viable successor to an old theory T_1 must "contain the laws of T as a limiting case."[24] John Watkins, a like-minded convergentist, puts the point this way:

It typically happens in the history of science that when some hitherto dominant theory T is superceded by T^1, T^1 is in the relation of correspondence to T [i.e., T is a 'limiting case' of T^1].[25]

Numerous recent philosophers of science have subscribed to a similar view, including Popper, H. R. Post, Krajewski, and N. Koertge.[26]

This form of retention is not the only one to have been widely discussed. Indeed, realists have espoused a wide variety of claims about what is or should be retained in the transition from a once successful predecessor (T_1) to a successor theory (T_2). Among the more important forms of realist retention are the following cases: (1) T_2 entails T_1 (W. Whewell); (2) T_2 retains the true consequences or truth content of T_1 (Popper); (3) T_2 retains the "confirmed" portions of T_1 (Post, Koertge); (4) T_2 preserves the theoretical laws and mechanisms of T_1 (Boyd, McMullin, Putnam); (5) T_2 preserves T_1 as a limiting case (J. Watkins, Putnam, Krajewski); (6) T_2 explains why T_1 succeeded insofar as T_1 succeeded (W. Sellars); and (7) T_2 retains reference for the central terms of T_1 (Putnam, Boyd).

The question before us is whether, when retention is understood in *any* of these senses, the realist's theses about convergence and retention are correct.

DO SCIENTISTS ADOPT THE RETENTIONIST STRATEGY OF CER?

One part of the convergent realist's argument is a claim to the effect that scientists generally adopt the strategy of seeking to preserve earlier theories in later ones. As Putnam puts it:

preserving the *mechanisms* of the earlier theory as often as possible, which is what scientists try to do.... That scientists try to do this... is a fact, and that this strategy has led to important discoveries... is also a fact.[27]

In a similar vein, I. Szumilewicz (although not stressing realism) insists that many eminent scientists made it a main heuristic requirement of their research programs that a new theory stand in a relation of 'correspondence' with the theory it supersedes.[28] If Putnam and the other retentionists are right about the strategy that most scientists have adopted, we should expect to find the historical literature of science abundantly provided with proofs that later theories do indeed contain earlier theories as limiting cases, or outright rejections of later theories that fail to contain earlier theories. Except on rare occasions (coming primarily from the history of mechanics), one finds neither of these concerns prominent in the literature of science. For instance, to the best of my knowledge, literally no one criticized the wave theory of light because it did not preserve the theoretical mechanisms of the earlier corpuscular theory; no one faulted C. Lyell's uniformitarian geology on the grounds that it dispensed with several causal processes prominent in catastrophist geology; Darwin's theory was not criticized by most geologists for its failure to retain many of the mechanisms of Lamarckian evolutionary theory.

For all the realist's confident claims about the prevalence of a retentionist strategy in the sciences, I am aware of *no* historical studies that would sustain as a *general* claim his hypothesis about the evaluative strategies utilized in science. Moreover, insofar as Putnam and Boyd claim to be offering "an explanation of the retentionist behavior of scientists,"[29] they have the wrong explanandum, for if there is any widespread strategy in science, it is one that says, "accept an empirically successful theory, regardless of whether it contains the theoretical laws and mechanisms of its predecessors."[30] Indeed, one could take a leaf from the realist's (C2) and claim that the success of the strategy of assuming that earlier theories do not generally refer shows that it is true that earlier theories generally do not!

(One might note in passing how often, and on what evidence, realists imagine that they are speaking for the scientific majority. Putnam, for instance, claims that "realism is, so to speak, 'science's philosophy of science'" and that "science taken at 'face value' *implies* realism."[31] C. A. Hooker insists that to be a realist is to take science "seriously,"[32] as if to suggest that conventionalists, instrumentalists, and positivists such as Duhem, Poincaré, and Mach did not take science seriously. The willingness of some realists to attribute realist strategies to working scientists— on the strength of virtually no empirical research into the principles which *in fact* have governed scientific practice—raises doubts about the seriousness of their avowed commitment to the empirical character of epistemic claims.)

DO LATER THEORIES PRESERVE THE MECHANISMS, MODELS, AND LAWS OF EARLIER THEORIES?

Regardless of the explicit strategies to which scientists have subscribed, are Putnam and several other retentionists right that later theories "typically" entail earlier theories, and that "earlier theories are, very often, limiting cases of later theories?"[33] Unfortunately, answering this question is difficult, since 'typically' is one of those weasel words that allows for much hedging. I shall assume that Putnam and Watkins mean that "most of the time (or perhaps in most of the important cases) successor theories contain predecessor theories as limiting cases." So construed, the claim is patently false. Copernican astronomy did not retain all the key mechanisms of Ptolemaic astronomy (e.g., motion along an equant); Newton's physics did not retain all (or even most of) the theoretical laws of Cartesian mechanics, astronomy, and optics; Franklin's electrical theory did not contain its predecessor (J. A. Nollet's) as a limiting case. Relativistic physics did not retain the ether, nor the mechanisms associated with it; statistical mechanics does not incorporate all the mechanisms of thermodynamics; modern genetics does not have Darwinian pangenesis as a limiting case; the wave theory of light did not appropriate the mechanisms of corpuscular optics; modern embryology incorporates few of the mechanisms prominent in classical embryological theory. As I have shown elsewhere,[34] loss occurs at virtually every level: the confirmed predictions of earlier theories are sometimes not explained by later ones; even the 'observable' laws explained by earlier theories are not always retained, not even as limiting cases; theoretical processes and mechanisms of earlier theories are, as frequently as not, treated as flotsam.

The point is that some of the most important theoretical innovations have been due to a willingness of scientists to violate the cumulationist or retentionist constraint which realists enjoin 'mature' scientists to follow.

There is a deep reason why the convergent realist is wrong about these matters. It has to do, in part, with the role of ontological frameworks in science and with the nature of limiting case relations. As scientists use the term 'limiting case', T_1 can be a limiting case of T_2 only if *all* the variables (observable and theoretical) assigned a value in T_1 are assigned a value by T_2, and if the values assigned to every variable of T_1 are the same as, or very close to, the values T_2 assigns to the corresponding variable when certain initial and boundary conditions—consistent with T_2[35]—are specified. This seems to require that T_1 can be a limiting case of T_2 only if *all* the entities postulated by T_1 occur in the ontology of T_2. Whenever there is a change of ontology accompanying a theory transition such that T_2 (when conjoined with suitable initial and boundary conditions) fails to capture the ontology of T_1, then T_1 cannot be a limiting case of T_2. Even where the ontologies of T_1 and T_2 overlap appropriately (i.e., where T_2's ontology embraces all of T_1's), T_1 is a limiting case of T_2 only if *all* the laws of T_1 can be derived from T_2, given appropriate limiting conditions. It is important to stress that *both* these conditions (among others) must be satisfied before one theory can be a limiting case of another. Where "closet" positivists might be content with capturing only the formal mathematical relations or only the observable consequences of T_1 within a successor T_2, any genuine realist must insist that T_1's underlying ontology is preserved in T_2's, *for it is that ontology above all which he alleges to be approximately true.*

Too often, philosophers (and physicists) infer the existence of a limiting-case relation between T_1 and T_2 on substantially less than this. For instance, many writers have claimed one theory to be a limiting case of another when only some, but not all, of the laws of the former are derivable from the latter. In other cases, one theory has been said to be a limiting case of a successor when the mathematical laws of the former find homologies in the latter but where the former's ontology is not fully extractable from the latter's.

Consider one prominent example which has often been misdescribed, namely, the transition from the classical ether theory to relativistic and quantum mechanics. It can, of course, be shown that *some* laws of classical mechanics are limiting cases of relativistic mechanics. But there are other laws and general assertions made by the classical theory (e.g., claims about the density and fine structure of the ether, general laws about the character of the interaction between ether and matter, models and mechanisms detailing the compressibility of the ether) which could not conceivably be limiting cases of modern mechanics. The reason is a simple one: a theory cannot assign values to a variable that does not occur in that theory's language (or, more colloquially, it cannot assign properties to entities whose existence it does not countenance). Classical ether physics contained a number of postulated mechanisms for dealing inter alia with the transmission of light through the ether. Such mechanisms could not possibly appear in a successor theory like the special theory of relativity which denies the very existence of an etherial medium and which accomplishes the explanatory tasks performed by the ether via very different mechanisms.

Nineteenth-century mathematical physics is replete with similar examples of evidently successful mathematical theories which, because some of their variables refer to entities whose existence we now deny, cannot be shown to be limiting cases of our physics. As Adolf Grünbaum has cogently argued, when we are confronted with two incompatible theories, T_1 and T_2, such that T_2 does not "contain" all of T_1's ontology, then not all the mechanisms and theoretical laws of T_1 that involve those entities of T_1 not postulated by T_2 can possibly be retained—not even as limiting cases—in T_2.[36] This result is of some significance. What little plausibility convergent or retentive realism has enjoyed derives from the presumption that it correctly describes the relationship between classical and postclassical mechanics and gravitational theory. Once we see that even in this prima facie most favorable case for the realist (where *some* of the laws of the predecessor theory are genuinely limiting cases of the successor), changing ontologies or conceptual frameworks make it impossible to capture many of the central theoretical laws and mechanisms postulated by the earlier theory, then we can see how misleading Putnam's claim is that "what scientists try to do [is to preserve] the *mechanisms* of the earlier theory as often as possible—or to show that they are 'limiting cases' of new mechanisms."[37] Where the mechanisms of the earlier theory involve entities whose existence the later theory denies, no scientist does (or should) feel any compunction about wholesale repudiation of the earlier mechanisms.

But even where there is no change in basic ontology, many theories (even in mature sciences such as physics) fail to retain all the explanatory successes of their predecessors. It is well known that statistical mechanics has yet to capture the irreversibility of macrothermodynamics as a genuine limiting case. Classical continuum mechanics has not yet been reduced to quantum mechanics or relativity. Contemporary field theory has yet to replicate the classical thesis that physical laws are invariant under reflection in space. If scientists had accepted the realist's constraint (namely, that new theories must have old theories as limiting cases), neither relativity nor statistical mechanics would have been viewed as viable theories. It has been said before, but it needs to be reiterated over and again: *a proof of the existence of limiting relations between selected components of two theories is a far cry from a systematic proof that one theory is a limiting case of the other.* Even if classical and modern physics stood to one another in the manner in which the convergent realist erroneously imagines they do, his hasty generalization that theory successions in all the advanced sciences show limiting-case relations is patently false.[38] But, as this discussion shows, not even the realist's paradigm case will sustain the claims he is apt to make about it.

What this analysis underscores is just how reactionary many forms of convergent epistemological realism are. If one took seriously CER's advice to reject any new theory that did not capture existing mature theories as referential and existing laws and mechanisms as approximately authentic, then any prospect for deep-structure, ontological changes in our theories would be foreclosed. Equally outlawed would be any significant repudiation of our theoretical models. In spite of his commitment to the growth of knowledge, the realist would unwittingly freeze science in its present state by forcing all future theories to accommodate the ontology of contemporary (mature) science and by foreclosing the possibility that some future generation may come to the conclusion that some (or even most) of the central terms in our best theories are no more referential than was 'natural place', 'phlogiston', 'ether', or 'caloric'.

COULD THEORIES CONVERGE IN WAYS REQUIRED BY THE REALIST?

These violations, within genuine science, of the sorts of continuity usually required by realists are by themselves sufficient to show that the form of scientific growth which the convergent realist takes as his explicandum is often absent, even in the mature sciences. But we can move beyond these specific cases to show in principle that the kind of cumulation demanded by the realist is unattainable. Specifically, by drawing on some results established by David Miller and others, the following can be shown:

1. The familiar requirement that a successor theory, T_2, must both preserve as true the true consequences of its predecessor, T_1, and explain T_1's anomalies is contradictory.

2. If a new theory, T_2, involves a change in the ontology or conceptual framework of a predecessor, T_1, then T_1 will have true and determinate consequences not possessed by T_2.

3. If two theories, T_1 and T_2, disagree, then each will have true and determinate consequences not exhibited by the other.

To establish these conclusions, one needs to utilize a "syntactic" view of theories according to which a theory is a conjunction of statements and its consequences are defined à la Tarski in terms of content classes. Needless to say, this is neither the only, nor necessarily the best, way of thinking about theories; but it happens to be the way in which most philosophers who argue for convergence and retention (e.g., Popper, Watkins, Post, Krajewski, and I. Niiniluoto) tend to conceive of theories. What can be said is that if one utilizes the Tarskian conception of a theory's content and its consequences as they do, then the familiar convergentist theses alluded to in conclusions 1 through 3 make no sense.

The elementary but devastating consequences of Miller's analysis establish that virtually any effort to link scientific progress or growth to the wholesale retention of a predecessor theory's Tarskian content *or* logical consequences *or* true consequences *or* observed consequences *or* confirmed consequences, is evidently doomed. Realists have not only got their history wrong insofar as they imagine that cumulative retention has prevailed in science, but we can also see that, given their views on what should be retained through theory change, history could not possibly have been the way their models require it to be. The realist's strictures on cumulativity are as ill advised normatively as they are false historically.

Along with many other realists, Putnam has claimed that "the mature sciences do converge . . . and that that convergence has great explanatory value for the theory of science."[39] As this section should show, Putnam and his fellow realists are arguably wrong on *both* counts. Popper once remarked that "no theory of knowledge should attempt to explain why we are successful in our attempts to explain things."[40] Such a dogma is too strong. But what the foregoing analysis shows is that an occupational hazard of recent epistemology is imagining that convincing explanations of our success come easily or cheaply.

SHOULD NEW THEORIES EXPLAIN WHY THEIR PREDECESSORS WERE SUCCESSFUL?

An apparently more modest realism than that outlined above is familiar in the form of the requirement (R4) often attributed to Sellars—that every satisfactory new theory must be able to explain why its predecessor was successful insofar as it was successful. On this view, viable new theories need not preserve all the content of their predecessors, nor capture those predecessors as limiting cases. Rather, it is simply insisted that a viable new theory, T_N, must explain why, when we conceive of the world according to the old theory T_O, there is a range of cases where our T_O-guided expectations are correct or approximately correct.

What are we to make of this requirement? In the first place, it is clearly *gratuitous*. If T_N has more confirmed consequences (and greater conceptual simplicity) than T_O, then T_N is preferable to T_O even if T_N cannot explain why T_O is successful. Contrariwise, if T_N has fewer confirmed consequences than T_O, then T_N cannot be rationally preferred to T_O even if T_N explains why T_O is successful. In short, a theory's ability to explain why a rival is successful is neither a necessary nor a sufficient condition for saying that it is better than its rival.

Other difficulties likewise confront the claim that new theories should explain why their predecessors were successful. Chief among them is the ambiguity of the notion itself. One way to show that an older theory, T_O, was successful is to show that it shares many confirmed consequences with a newer theory, T_N, which is highly successful. But this is not an "explanation" that a scientific realist could accept, since it makes no reference to, and thus does not depend upon, an epistemic assessment of either T_O or T_N. (After all, an instrumentalist could quite happily grant that if T_N "saves the phenomena" then T_O—insofar as some of its observable consequences overlap with or are experimentally indistinguishable from those of T_N—should also succeed at saving the phenomena.)

The intuition being traded on in this persuasive account is that the pragmatic success of a new theory, combined with a partial comparison of the respective consequences of the new theory and its predecessor, will sometimes put us in a position to say when the older theory worked and when it failed. But such comparisons as can be made in this manner do not involve *epistemic* appraisals of either the new or the old theory qua theories. Accordingly, the possibility of such comparisons provides no argument for epistemic realism.

What the realist apparently needs is an epistemically robust sense of "explaining the success of a predecessor." Such an epistemic characterization would presumably begin with the claim that T_N, the new theory, was approximately true and would proceed to show that the 'observable' claims of its predecessor, T_O, deviated only slightly from (some of) the 'observable' consequences of T_N. It would then be alleged that the (presumed) approximate truth of T_N and the partially overlapping consequences of T_O and T_N jointly explained why T_O was successful insofar as it was successful. But this is a nonsequitur. As I have shown above, the fact that a T_N is approximately true does not even explain why it is successful; how, under those circumstances, can the approximate truth of T_N explain why some theory different from T_N is successful? Whatever the nature of the relations between T_N and T_O (entailment, limiting case, etc.), the epistemic ascription of approximate truth to either T_O or T_N (or both) apparently leaves untouched questions of how successful T_O or T_N are.

The idea that new theories should explain why older theories were successful (insofar as they were) originally arose as a rival to the "levels" picture of explanation according to which new theories fully explained, because they entailed, their predecessors. It is clearly an improvement over the levels picture (for it does recognize that later theories generally do not entail their predecessors). But when it is formulated as a general thesis about intertheory relations, designed to buttress a realist epistemology, it is difficult to see how this position avoids difficulties similar to those discussed in earlier sections.

THE REALISTS' ULTIMATE *PETITIO PRINCIPII*

It is time to step back a moment from the details of the realists' argument to look at its general strategy. Fundamentally, the realist is utilizing, as we have seen, an abductive inference which proceeds from the success of science to the conclusion that science is approximately true, verisimilar, or referential (or any combination of these). This argument is meant to show the skeptic that theories are not ill gotten, the positivist that theories are not reducible to their observational consequences, and the pragmatist that classical epistemic categories (e.g., "truth" and "falsehood") are a relevant part of metascientific discourse.

It is little short of remarkable that realists would imagine that their critics would find the argument compelling. As I have shown elsewhere,[41] ever since antiquity critics of epistemic realism have based their skepticism upon a deep-rooted conviction that the fallacy of affirming the consequent is indeed fallacious. When E. Sextus or R. Bellarmine or Hume doubted that certain theories which saved the phenomena were warrantable as true, their doubts were based on a belief that the exhibition that a theory had some true consequences left entirely open the truth-status of the theory. Indeed, many nonrealists have been nonrealists precisely because they believed that false theories, as well as true ones, could have true consequences.

Now enters the new breed of realist (e.g., Putnam, Boyd, Newton-Smith) who wants to argue that epistemic realism can reasonably be presumed true by virtue of the fact that it has true consequences. But this is a monumental case of begging the question. The nonrealist refuses to admit that a *scientific* theory can be warrantedly judged to be true simply because it has some true consequences. Such nonrealists are not likely to be impressed by the claim that a philosophical theory such as realism can be warranted as true because it arguably has some true consequences. If nonrealists are chary about first-order abductions to avowedly true conclusions, they are not likely to be impressed by second-order abductions, particularly when, as I have tried to show above, the premises and conclusions are so indeterminate.

But, it might be argued, the realist is not out to convert the intransigent skeptic or the determined instrumentalist.[42] Perhaps, he is seeking to show that realism can be tested like any other scientific hypothesis, and that realism is at least as well confirmed as some of our best scientific theories. Such an analysis, however plausible initially, will not stand up to scrutiny. I am aware of no realist who is willing to say that a scientific theory can be reasonably presumed to be true or even regarded as well confirmed just on the strength of the fact that its thus-far-tested consequences are true. Realists have long been in the forefront of those opposed to ad hoc and post hoc theories. Before a realist accepts a scientific hypothesis, he generally wants to know whether it has explained or predicted more than it was devised to explain, whether it has been subjected to a battery of controlled tests, whether it has successfully made novel predictions, and whether there is independent evidence for it.

What, then, of realism itself as a "scientific" hypothesis?[43] Even if we grant (contrary to what I argued in the section on "Approximate Truth and Success") that realism entails and thus explains the success of science, ought that (hypothetical) success warrant, by the realist's own construal of scientific acceptability, the acceptance of realism? Since realism was devised to explain the success of science, it remains purely ad hoc with respect to that success. If realism has made some novel predictions or has been subjected to carefully controlled tests, one does not learn about it from the literature of contemporary realism. At the risk of apparent inconsistency, the realist repudiates the instrumentalist's view that saving the phenomena is a significant form of evidential support while endorsing realism itself on the transparently instrumentalist grounds that it is confirmed by those very facts it was invented to explain. No proponent of realism has sought to show that realism satisfies those stringent empirical demands which the realist himself minimally insists on when appraising scientific theories. The latter-day realist often calls realism a "scientific" or "well-tested" hypothesis but seems curiously reluctant to subject it to those controls he otherwise takes to be a sine qua non for empirical well-foundedness.

CONCLUSION

The arguments and cases discussed above seem to warrant the following conclusions:

1. The fact that a theory's central terms refer does not entail that it will be successful, and a theory's success is no warrant for the claim that all or most of its central terms refer.

2. The notion of approximate truth is presently too vague to permit one to judge whether a theory consisting entirely of approximately true laws would be empirically successful. What is clear is that a theory may be empirically successful even if it is not approximately true.

3. Realists have no explanation whatever for the fact that many theories which are not approximately true and whose "theoretical" terms seemingly do not refer are, nonetheless, often successful.

4. The convergentist's assertion that scientists in a "mature" discipline usually preserve, or seek to preserve, the laws and mechanisms of earlier theories in later ones is probably false. His assertion that when such laws are preserved in a successful successor, we can explain the success of the latter by virtue of the truthlikeness of the preserved laws and mechanisms, suffers from all the defects noted above confronting approximate truth.

5. Even if it could be shown that referring theories and approximately true theories would be successful, the realist's argument that successful theories are approximately true and genuinely referential takes for granted precisely what the nonrealist denies, namely, that explanatory success betokens truth.

6. It is not clear that acceptable theories either *do* or *should* explain why their predecessors succeeded or failed. If a theory is better supported than its rivals and predecessors, then it is not epistemically decisive whether it explains why its rivals worked.

7. If a theory has once been falsified, it is unreasonable to expect that a successor should retain either all of its content *or* its confirmed consequences *or* its theoretical mechanisms.

8. Nowhere has the realist established, except by fiat, that nonrealist epistemologists lack the resources to explain the success of science.

With these specific conclusions in mind, we can proceed to a more global one: it is not yet established—Putnam, Newton-Smith, and Boyd notwithstanding—that realism can explain *any* part of the successes of science. What is very clear is that realism *cannot*, even by its own lights, explain the success of those many theories whose central terms have evidently not referred and whose theoretical laws and mechanisms were not approximately true. The inescapable conclusion is that insofar as many realists are concerned with explaining how science works and with assessing the adequacy of their epistemology by that standard, they have, thus far, failed to explain very much. Their epistemology is confronted by anomalies that seem beyond its resources to grapple with.

It is important to guard against a possible misinterpretation of this essay. Nothing I have said here refutes the possibility, in principle, of a realistic epistemology of science. To conclude as much would be to fall prey to the same inferential prematurity with which many realists have rejected in principle the possibility of explaining science in a nonrealist way. My task here is, rather, that of reminding ourselves that there *is* a difference between wanting to believe something and having good reasons for believing it. All of us would like realism to be true; we would like to think that science works because it has got a grip on how things really are. But such claims have yet to be made out. Given the present state of the art, it can only be wish fulfillment that gives rise to the claim that realism, and realism alone, explains why science works.

NOTES

1. R. Boyd, "Realism, Underdetermination, and a Causal Theory of Evidence," *Nous* 7 (1973): 1-12; W. Newton-Smith, "The Underdetermination of Theories by Data," *Proceedings of the Aristotelian Society* (1978), 71-91; H. Putnam, *Mathematics, Matter, and Method*, vol. 1 (Cambridge: Cambridge University Press, 1975); Ilkka Niiniluoto, "On the Truthlikeness of Generalizations," in *Basic Problems in Methodology and Linguistics*, ed. R. Butts and J. Hintikka (Dordrecht: D. Reidel, 1977), 121-147.

2. See Putnam's essay in this volume (ed.).

3. Putnam, evidently following Body, sums up theses R1 and R3 in these words:

"1) Terms in a mature science typically *refer*.

2) The laws of a theory belonging to a mature science are typically approximately true.

...I will only consider [new] theories...which have this property—[they] contain the [theoretical] laws of [their predecessors] as a limiting case."

H. Putnam, *Meaning and the Moral Sciences* (London: Routledge and Kegan Paul, 1978), 20-21.

4. Putnam insists, for instance, that if the realist is wrong about theories being referential, then "the success of science is a miracle" (Putnam, *Mathematics, Matter, and Method*, 1: 69).

5. Boyd remarks: "Scientific realism offers an *explanation* for the legitimacy of ontological commitment to theoretical entities" (Putnam, *Meaning and the Moral Sciences*, 2). It allegedly does so by explaining why theories containing theoretical entities work so well: because such entities genuinely exist.

6. Whether one utilizes Putnam's earlier or later versions of realism is irrelevant for the central arguments of this essay.

7. Putnam, *Meaning and the Moral Sciences*, 20-22.

8. Ibid., 22.

9. Ibid., 20.

10. "Realism, Underdeterminism, and a Causal Theory of Evidence," 1. See also p. 3: "experimental evidence for a theory is evidence for the truth of even its non-observational laws." See also W. Sellars, *Science, Perception and Reality* (New York: The Humanities Press, 1963), 97.

11. A caveat is in order here. *Even* if all the central terms in some theory refer, it is not obvious that every rational successor to that theory must preserve all the referring terms of its predecessor. One can easily imagine circumstances when the new theory is preferable to the old one even though the range of application of the new theory is less broad than the old. When the range is so restricted, it may well be entirely appropriate to drop reference to some of the entities that figured in the earlier theory.

12. For Putnam and Boyd both, "it will be a constraint on T_2 [i.e., any new theory in a domain]... that T_2 must have this property, the property that *from its standpoint* one can assign referents to the terms of T_1 [i.e., an earlier theory in the same domain]" (Putnam, "What is Realism?" this volume). For Boyd, see "Realism, Underdeterminism, and a Causal Theory of Evidence," 8: "new theories should, *prima facie*, resemble current theories with respect to their accounts of causal relations among theoretical entities."

13. Putnam, *Mathematics, Matter, and Method*, 22.

14. For just a small sampling of this view, consider the following: "the claim of a realist ontology of science is that the only way of explaining why the models of science function so successfully... is that they approximate in some way the structure of the object" (Ernan McMullin, "The History and Philosophy of Science: A Taxonomy," in *Minnesota Studies in the Philosophy of Science*, ed. R. Stuewer, 5 [1970]: 63-64); "the continued success of confirmed theories can be *explained* by the hypothesis that they are in fact close to the truth" (Niiniluoto); and the claim that "the laws of a theory belonging to a mature science are typically approximately *true*... [provides] an explanation of the behavior of scientists and the success of science" (Putnam, *Meaning and the Moral Sciences*, 20-21). J. J. Smart, W. Sellars, and Newton-Smith, among others, share a similar view.

15. Although Popper is generally careful not to assert that actual historical theories exhibit ever-increasing truth content (for an exception, see his *Conjectures and Refutations* [London: Routledge and Kegan Paul, 1963], 220), other writers have been more bold. Thus, Newton-Smith writes that "the historically generated sequence of theories of a mature science is a sequence in which succeeding theories are increasing in truth content without increasing in falsity content" (W. Newton-Smith, "In Defense of Truth," forthcoming).

16. On the more technical side, Niiniluoto has shown that a theory's degree of corroboration covaries with its "estimated verisimilitude" (Niiniluoto, "On the Truthlikeness of Generalizations," 121-147). Roughly speaking, "estimated truthlikeness" is a measure of how closely (the content of) a theory corresponds to *what we take to be* the best conceptual systems that, so far, we have been able to find (Ilkka Niiniluoto, "Scientific Progress," *Synthese* 45 (1980): 443 ff.). If Niiniluoto's measures work, it follows from the above-mentioned covariance that an

empirically successful theory will have a high degree of estimated truthlikeness. But because estimated truthlikeness and genuine verisimilitude are not necessarily related (the former being parasitic on existing evidence and available conceptual systems), it is an open question whether, as Niiniluoto asserts, the continued success of highly confirmed theories can be *explained* by the hypothesis that they in fact are so close to the truth, at least in the relevant respects. Unless I am mistaken, this remark of his betrays a confusion between "true verisimilitude" (to which we have no epistemic access) and "estimated verisimilitude" (which is accessible but nonepistemic).

17. Newton-Smith, "In Defense of Truth," 16.

18. Newton-Smith claims that the increasing predictive success of science through time "would be totally mystifying . . . if it were not for the fact that theories are capturing more and more truth about the world" (Newton-Smith, "In Defense of Truth," 15).

19. I must stress again that I am *not* denying that there *may* be a connection between approximate truth and predictive success. I am only observing that until the realists show us what that connection is, they should be more reticent than they are about claiming that realism can explain the success of science.

20. A *nonrealist* might argue that a theory is approximately true just in case all its *observable* consequences are true or within a specified interval from the true value. Theories that were "approximately true" in this sense would indeed be demonstrably successful. But, the realist's (otherwise commendable) commitment to taking seriously the theoretical claims of a theory precludes him from utilizing any such construal of approximate truth, since he wants to say that the theoretical as well as the observational consequences are approximately true.

21. W. Krajewski, *Correspondence Principle and Growth of Science* (Dordrecht: D. Reidel, 1977), 91.

22. Putnam, *Meaning and the Moral Sciences*, 21.

23. If this argument, which I attribute to the realists, seems a bit murky, I challenge any reader to find a more clear-cut one in the literature! Overt formulations of this position can be found in Putnam, Boyd, and Newton-Smith.

24. Putnam, *Meaning and the Moral Sciences*, 21.

25. John Watkins, "Corroboration and the Problem of Content-Comparison," in *Progress and Rationality in Science*, ed. G. Radnitzky and G. Anderson (Dordrecht: D. Reidel, 1978), 376-377.

26. Popper: "A theory which has been well-corroborated can only be superseded by one . . . [which] *contains* the old well-corroborated theory—or at least a good approximation to it." K. Popper, *Logic of Scientific Discovery* (New York: Basic Books, 1959), 276.

Post: "I shall even claim that, as a matter of empirical historical fact, [successor] theories [have] always explained the *whole* of [the well-confirmed part of their predecessors]." H. R. Post, "Correspondence, Invariance and Heuristics: In Praise of Conservative Induction," *Studies in the History and Philosophy of Science* 2 (1971): 229.

Koertge: "Nearly all parts of successive theories in the history of science stand in a correspondence relation and . . . where there is no correspondence to begin with, the new theory will be developed in such a way that it comes more nearly into correspondence with the old." N. Koertge, "Theory Change in Science," in *Conceptual Change*, ed. G. Pearce and P. Maynard (Dordrecht: D. Reidel, 1973), 176-177.

Among other authors who have defended a similar view, one should mention A. Fine, "Consistency, Derivability and Scientific Change," *Journal of Philosophy* 64 (1967): 231 ff.; C. Kordig, "Scientific Transitions, Meaning Invariance, and Derivability," *Southern Journal of Philosophy* (1971): 119-125; H. Margenau, *The Nature of Physical Reality* (New York: McGraw-Hill, 1950); and L. Sklar, "Types of Inter-Theoretic Reductions," *British Journal for Philosophy of Science* 18 (1967): 190-224.

27. Putnam fails to point out that it is also a fact that many scientists do *not* seek to preserve earlier theoretical mechanisms and that theories which have not preserved earlier theoretical mechanisms (whether the germ theory of disease, plate tectonics, or wave optics) have led to important discoveries is also a fact.

28. I. Szumilewicz, "Incommensurability and the Rationality of the Development of Science," *British Journal for the Philosophy of Science* 28 (1977): 348.

29. Putnam, *Meaning and the Moral Sciences*, 21.

30. I have written a book about this strategy. See Larry Laudan, *Progress and Its Problems* (Berkeley, Los Angeles, London: University of California Press, 1977).

31. After the epistemological and methodological battles about science during the last three hundred years, it should be fairly clear that science, taken at its face value, implies no particular epistemology.

32. Clifford Hooker, "Systematic Realism," *Synthese* 26 (1974): 467-472.

33. Putnam, *Meaning and the Moral Sciences*, 20, 123.

34. Larry Laudan, "Two Dogmas of Methodology," *Philosophy of Science* 43 (1976): 467-472.

35. This matter of limiting conditions consistent with the "reducing" theory is curious. Some of the best-known expositions of limiting-case relations depend (as Krajewski has observed) upon showing an earlier theory to be a limiting case of a later theory only by adopting limiting assumptions *explicitly denied by the later theory*. For instance, several standard textbook discussions present (a portion of) classical mechanics as a limiting case of special relativity, provided c approaches infinity. But special relativity is committed to the claim that c is a constant. Is there not something suspicious about a "derivation" of T_1 from a T_2 which essentially involves an assumption inconsistent with T_2? If T_2 is correct, then it forbids the adoption of a premise commonly used to derive T_1 as a limiting case. (It should be noted that most such proofs can be reformulated unobjectionably, e.g., in the relativity case, by letting $v \to 0$ rather than $v \to \infty$.)

36. Adolf Grünbaum, "Can a Theory Answer More Questions than One of Its Rivals?" *British Journal for Philosophy of Science* 27 (1976): 1-23.

37. Putnam, *Meaning and the Moral Sciences*, 20.

38. As Mario Bunge has cogently put it: "The popular view on inter-theory relations... that every new theory includes (as regards its extension) its predecessors... is philosophically superficial... and it is false as a historical hypothesis

concerning the advancement of science." M. Bunge, "Problems Concerning Inter-theory Relations," in *Induction, Physics and Ethics*, ed. P. Weingartner and G. Zecha (Dordrecht: D. Reidel, 1970), 309-310.

39. Putnam, *Meaning and the Moral Sciences*, 37.

40. K. Popper, *Conjectures and Refutations* (London: Routledge and Kegan Paul, 1963), Introduction.

41. Larry Laudan, "Ex-Huming Hacking," *Erkenntis* 13 (1978).

42. I owe the suggestion of this realist response to Andrew Lugg.

43. I find Putnam's views on the "empirical" or "scientific" character of realism rather perplexing. At some points, he seems to suggest that realism is both empirical and scientific. Thus, he writes: "If realism is an explanation of this fact [namely, that science is successful], realism must itself be an over-arching scientific *hypothesis*" ("What is Realism?" this volume). Since Putnam clearly maintains the antecedent, he seems committed to the consequent. Elsewhere he refers to certain realist tenets as being "our highest level empirical generalizations about knowledge" (*Meaning and the Moral Sciences*, 37). He says, moreover, that realism "could be false," and that "facts are relevant to its support (or to criticize it)" (pp. 78-79). Nonetheless, for reasons he has not made clear, Putnam wants to deny that realism is either scientific or a hypothesis (p. 79). How realism can consist of doctrines which explain facts about the world, are empirical generalizations about knowledge, and can be confirmed or falsified by evidence, and *yet* be neither scientific nor hypothetical, is left opaque.

12

To Save the Phenomena

Bas C. van Fraassen

After the demise of logical positivism, scientific realism has once more returned as a major philosophical position. I shall not try here to criticize that position, but rather attempt to outline a comprehensive alternative.[1]

I

What exactly is scientific realism? Naively stated, it is the view that the picture science gives us of the world is true, and that the entities postulated really exist. (Historically, it added that there are real necessities in nature; I will ignore that aspect here.)[2] But that statement is too naive; it attributes to the scientific realist the belief that today's scientific theories are (essentially) right.

The correct statement, it seems to me, must indeed be in terms of epistemic attitudes, but not so directly. The aim of science is to give us *a literally true story of what the world is like*; and the proper form of acceptance of a theory is to believe that it is true. This is the statement of scientific realism: "To have good reason to accept a theory is to have good reason to believe that the entities it postulates are real," as Wilfrid Sellars has expressed it. Accordingly, an antirealism is a position according to which the aims of science can well be served without giving such a literally true story, and acceptance of a theory may properly involve something less (or other) than belief that it is true.

The idea of a literally true account has two aspects: the language is to be literally construed; and, so construed, the account is true. This divides the antirealists into two sorts. The first sort holds that science is or aims to be true, properly (but not literally) construed. The second holds that the language of science should be literally construed, but its theories need not be true to be good. The antirealism I advocate belongs to the second sort.

II

When Newton wrote his *Mathematical Principles of Natural Philosophy* and *System of the World,* he carefully distinguished the phenomena to be saved from the reality he postulated. He distinguished the "absolute magnitudes" that appear in his axioms from their "sensible measures" which are determined experimentally. He discussed carefully the ways in which, and the extent to which, "the true motions of particular bodies [may be determined] from the apparent," via the assertion that "the apparent motions...are the differences of true motions."[3]

The "apparent motions" form relational structures defined by measuring relative distances, time intervals, and angles of separation. For brevity, let us call these relational structures *appearances.* In the mathematical model provided by Newton's theory, bodies are located in absolute space, in which they have real or absolute motions. But within these models we can define structures that are meant to be exact reflections of those appearances and are, as Newton says, identifiable as differences between true motions. These structures, defined in terms of the relevant relations between absolute locations and absolute times, which are the appropriate parts of Newton's models, I will call *motions,* borrowing Herbert Simon's term.[4]

When Newton claims empirical adequacy for his theory, he is claiming that his theory has some model such that *all actual appearances are identifiable with (isomorphic to) motions in that model.*

Newton's theory goes a great deal further than this. It is part of his theory that there is such a thing as absolute space, that absolute motion is motion relative to absolute space, that absolute acceleration causes certain stresses and strains and thereby deformations in the appearances, and so on. He offered, in addition, the "hypothesis" (his term) that the center of gravity of the solar system is at rest in absolute space. But, as he himself noted, the appearances would be no different if that center were in any other state of constant absolute motion.

Let us call Newton's theory (mechanics and gravitation) TN, and $TN(v)$ the theory TN plus the postulate that the center of gravity of the solar system has constant absolute velocity v. By Newton's own account, he claims empirical adequacy for $TN(O)$; and also claims that if $TN(O)$ is empirically adequate, then so are all the theories $TN(v)$.

Recalling what it was to claim empirical adequacy, we see that all the theories $TN(v)$ are empirically equivalent exactly if all the motions in a model of $TN(v)$ *are isomorphic to motions in a model* $TN(v+w)$, for all constant velocities v and w. For now, let us agree that these theories are empirically equivalent, referring objections to a later section.

III

What exactly is the empirical import of $TN(O)$: Let us focus on a fictitious and anachronistic philosopher Leibniz*, whose only quarrel with Newton's theory is that he does not believe in the existence of absolute space. As a corollary, of course, he can attach no physical significance to statements about absolute motion. Leibniz* believes, like Newton, that $TN(O)$ is empirically adequate; but not that it is true. For the sake of brevity, let us say that Leibniz* *accepts* the theory but that he does not *believe* it; when confusion threatens we may expand that idiom to say that he *accepts the theory as empirically adequate* but does not *believe it to be true.* What does Leibniz* believe, then?

Leibniz* believes that $TN(O)$ is empirically adequate, and hence, equivalently, that all the theories $TN(v)$ are empirically adequate. Yet we cannot identify the theory that Leibniz* holds about the world—call it TNE—with the common part of all the theories $TN(v)$. For each of these theories $TN(v)$ has such consequences as that the earth has *some* absolute velocity, and that absolute space exists. In each model of theory $TN(v)$ there is to be found something other than motions, and there is the rub.

To believe a theory is to believe that one of its models correctly represents the world. A theory may have isomorphic models; that redundancy is easily removed. If it has been removed, then to believe the theory is to believe that exactly one of its models correctly represents the world. Therefore, if we believe of a family of theories that all are empirically adequate, but each goes beyond the phenomena, then we are still free to believe that each is false and, hence, their common part is false. For that common part is phrasable as: one of the models of one of those theories correctly represents the world.

IV

It may be objected that theories will seem empirically equivalent only so long as we do not consider their possible extensions.[5] The equivalence may generally, or always, disappear when we consider their implications for some further domain of application. The usual example is Brownian motion; but this is imperfect, for it was known that phenomenological and statistical thermodynamics disagreed even on macroscopic phenomena over sufficiently long periods of time. But there is a good, *fictional* example; the combination of electromagnetism with mechanics, if we ignore the unexpected null results that led to the replacement of classical mechanics.

Maxwell's theory was not developed as part of mechanics, but it did have mechanical models. This follows from a result of J. Koenig, as detailed by Poincaré in the preface of his *Electricité et Optique* and elsewhere. But the theory had the strange new feature that velocity itself, not just its derivative, appears in the equations. A spate of thought experiments was designed to measure absolute velocity, the simplest perhaps that of Poincaré:

> Consider two electrified bodies; though they seem to us as at rest, they are both carried along by the motion of the earth; . . . therefore, equivalent to two parallel currents of the same sense and these two currents should attract each other. In measuring this attraction, we shall measure the velocity of the earth; not its velocity in relation to the sun or the fixed stars, but its absolute velocity.[6]

The null outcome of all experiments of this sort led to the replacement of classical by relativistic mechanics. But let us imagine that values *were* found for the absolute velocities; specifically, for that of the center of the solar system. Then, surely, would one of the theories $TN(v)$ be confirmed and the others falsified?

This reasoning is spurious. Newton made the distinction between true and apparent motions without presupposing more than the basic mechanics in which Maxwell's theory has models. Each motion in a model of $TN(v)$ is isomorphic to one in some model of $TN(v+w)$, for all constant velocities v and w. Could this assertion of empirical equivalence possibly be controverted by those nineteenth-century reflections? The answer is "no." The thought experiment, we may imagine, confirmed the theory that added to TN the hypotheses:

HO: The center of gravity of the solar system is at absolute rest.
EO: Two electrified bodies moving with absolute velocity v attract each other with force $F(v)$.

This theory has a consequence strictly about appearances:

CON: Two electrified bodies, moving with velocity v relative to the center of gravity of the solar system, attract each other with force $F(v)$.

However, that same consequence can be had by adding to TN the two alternative hypotheses:

Hw: The center of gravity of the solar system has absolute velocity *w*.

Ew: Two electrified bodies moving with absolute velocity $v+w$ attract each other with force $F(v)$.

More generally, for each theory $TN(v)$, there is an electromagnetic theory $E(v)$ such that $E(O)$ is Maxwell's and all the combined theories $TN(v)$ plus $E(v)$ are empirically equivalent.

There is no originality in this observation, of which Poincaré discusses the equivalent immediately after the passage I cited above. It seems that only familiar examples, but rightly stated, are needed to show the feasibility of concepts of empirical adequacy and equivalence. In the remainder of this paper I will try to generalize these considerations, while showing that the attempts to explicate those concepts *syntactically* had to reduce them to absurdity.

V

The idea that theories may have hidden virtues by allowing successful extensions to new kinds of phenomena is too pretty to be left—nor is it a very new idea. In the first lecture of his *Cours de philosophie positive*, Comte referred to Fourier's theory of heat as showing the emptiness of the debate between partisans of calorific matter and kinetic theory. The illustrations of empirical equivalence have that regrettable tendency to date; calorifics lost. Federico Enriques seemed to place his finger on the exact reason when he wrote: "The hypotheses which are indifferent in the limited sphere of the actual theories acquire significance from the point of view of their possible extension."[7] To evaluate this suggestion, we must ask what exactly is an extension of a theory.

Suppose that experiments really had confirmed the combined theory $TN(O)$ plus $E(O)$. In that case, mechanics would have won a victory. The claim that $TN(O)$ was empirically adequate would have been borne out by the facts. But such victorious extensions could never count for a theory as against one of its empirical equivalents.

Therefore, if Enriques's idea is to be correct, there must be another sort of extension, which is really a defeat—but a qualified defeat. For a theory *T* may have an easy or obvious modification that is empirically adequate, while another theory empirically equivalent to *T* does not. One example may be the superiority of Newton's celestial mechanics over the variant produced by Brian Ellis; Ellis himself seems to be of this opinion.[8] This is a *pragmatic* superiority and cannot suggest that theories, empirically equivalent in the sense explained, can nevertheless have different empirical import.

VI

We still need a general account of empirical adequacy and equivalence. It is here that the syntactic approach has most conspicuously failed. A theory was conceived as identifiable with the set of its theorems in a specified language. This language has a vocabulary, divided into two classes—the observational and theoretical terms. Let the first class be E; then the empirical import of theory T was said to be its subtheory T/E—those theorems expressible in that subvocabulary. T and T' were declared empirically equivalent if T/E was the same as T'/E.

Obvious questions were raised and settled. Craig showed that, under suitable conditions, T/E is axiomatizable in the vocabulary E. Logicians attached importance to questions about restricted vocabularies, and this was apparently enough to make philosophers think them important too. The distinction between observational and theoretical terms was more debatable, and some changed the division into "old" and "newly introduced" terms.[9] But all this is mistaken. Empirical import cannot be isolated in this syntactic fashion. If that could be done, then T/E would say exactly what T says about what is observable, and nothing else. But consider: the quantum theory, Copenhagen version, says that there are things which sometimes have a position in space and sometimes do not. This consequence I have just stated without using theoretical terms. Newton's theory TN implies that there is something (to wit, absolute space) that neither has a position nor occupies volume. As long as unobservable entities differ systematically from observable entities with respect to observable characteristics, T/E will say that there are such things if T does.

The reduced theory T/E is not a description of the observable part of the world of T; rather, it is a hobbled and hamstrung version of T's description of everything. Empirical equivalence fares as badly. In section II, $TN(O)$ and TNE *must* be empirically equivalent, but the above remark about TN shows that $TN(O)$ is not TNE/E. To eliminate such embarrassments, extensions of theories were considered in attempts to redefine empirical equivalence.[10] But these have similar absurd consequences.

The worst consequence of the syntactic approach was surely the way it focused philosophical attention on irrelevant technical questions. The expressions 'theoretical object' and 'observational predicate' mark category mistakes. Terms may be theoretical, but 'observable' classifies putative entities. Hence there cannot be a "theoretical/observable distinction." It is true surely that elimination of all theory-laden terms would leave no usable language; also that 'observable' is as vague as 'bald'. But these facts imply not at all that 'observable' marks an unreal distinction. It refers quite clearly to our limitations, the limits of observation, which are not incapacitating but also are not negligible.

VII

The phenomena are saved when they are exhibited as fragments of a larger unity. For that very reason it would be strange if scientific theories described the phenomena, the observable part, in different terms from the rest of the world they describe. And so an attempt to draw the conceptual line between phenomena and the transphenomenal by means of a distinction of vocabulary must always have looked too simple to be good.

Not all philosophers who discussed unobservables, by any means, did so in terms of vocabulary. But there was a common assumption: that the distinction marked is philosophical. Hence it must be drawn, if at all, by philosophical analysis and, if attacked, by philosophical arguments. This attitude needs a grand reversal. If there are limits to observation, these are empirical and must be described by empirical science. The classification marked by 'observable' must be of entities in the world of science. And science, in giving content to the distinction, will reveal how much we believe when we accept it as empirically adequate.

A future unified science may detail the limits of observation exactly; meanwhile, extant theories are not silent on them. We saw Newton's delineation; for relativity theory, we have two revealing studies by Clark Glymour. The first shows that local (hence, I should think, measurable) quantities do not uniquely determine global features of space-time.[11] The second shows that these features also are not uniquely determined by structures each lying wholly within some absolute past cone—hence, I should think, by observable structures. It is the theory of relativity itself, after all, that places an *absolute* limit on the information we can gather, through the limiting function of the speed of light.

In the foundations of quantum mechanics much more attention has been given to measurement. Much of the discussion is about necessary limitations: the role of noise in amplification, the distinction between macro- and micro-observables.[12] Yet we have no such clarity as Glymour gave us for relativity theory, concerning the extent to which macrostructure determines microstructure. The debate over scientific realism may at least have the virtue of directing attention to such questions.

Science itself distinguishes the observable that it postulates from the whole it postulates. The distinction, being in part a function of the limits science discloses on human observation, is anthropocentric. But, since science places human observers among the physical systems it means to describe, it also gives itself the task of describing anthropocentric distinctions. It is in this way that even the scientific realist must observe a distinction between the phenomena and the transphenomenal in the scientific world picture.

VIII

I have laid some philosophical misfortunes at the door of a mistaken orientation toward syntax. The alternative is to say that theories are presented directly by describing their models. But does this really introduce a new element? When you give the theorems of *T*, you give the set of models of *T*—namely, all those structures that satisfy the theorems. And if you give the models, you give at least the set of theorems of *T*—namely all those sentences that are satisfied in all the models. Does it not follow that we can as advantageously identify *T* with its theorems as with its models?

But there is an ellipsis in the argument. It is being assumed that there is a specific language *L* which is the one language that belongs to *T*. And indeed, the theorems of *T* in *L* determine and are determined by the set of model structures of *L* (that is, structures in which *L* is interpreted) in which those theorems are satisfied. However, the assumption that there is a language *L* which plays this role for *T* places important restrictions on what the set of models of *T* can be like.

A theory provides, among other things, a specification (more or less complete) of the parts of its models that are to be direct images of the structures described in measurement reports. In the case of Newton's mechanics, I called those parts "motions," in general, let us call them "empirical substructures." The structures described in measurement reports we may continue to call "appearances." A theory is *empirically adequate* exactly if all appearances are isomorphic to empirical substructures in at least one of its models. Theory *T* is *empirically no stronger* than theory *T'* exactly if, for each model *M* of *T*, there is a model *M'* of *T'* such that all empirical substructures of *M* are isomorphic to empirical substructures of *M'*. Theories *T* and *T'* are 'empirically equivalent' exactly if neither is empirically stronger than the other. In that case, as an easy corollary, each is empirically adequate if and only if the other is.

In section V, I distinguished two kinds of extensions, the first a sort of victory and the second a sort of defeat. Let us call the first a *proper extension*: this simply narrows the class of models. We may call a theory *empirically minimal* if it is not empirically equivalent to any of its proper extensions. Glymour has convincingly argued, in the work cited above, that general relativity is not empirically minimal. The reason is, in my present terms, that only local properties of space-time enter the descriptions of the appearances, but models may differ in global properties. This is a further nontrivial example of empirical equivalence.

The second sort of extension I shall not try to define precisely. The idea is that models of the theory may differ in structure other than that of the empirical substructures. In that case the theory is not empirically minimal, but this may put it in the advantageous position of offering modeling possibilities when radically new phenomena come to light. An example may yet be offered by hidden-variable theories in quantum mechanics.[13]

In terms of the concepts now at our disposal, and the examples given, we can conclude that there are indeed nontrivial cases of empirical equivalence, nonuniqueness, and extendability, both proper and improper. Such cases are now seen to be quite possible *even if the formulation of the theory has not a single term that cannot be called observational, in some way.* And now it should be possible to state the issue of scientific realism, which concerns our epistemic attitude toward theories rather than their internal structure.

Not all the results of measurements are in; they are never all in. Therefore, we cannot know what all the appearances are. We can say that a theory is empirically adequate, that all the appearances will fit (the empirical substructures of) its models. Though we cannot know this with certainty, we can reasonably believe it. All this is the case not only for empirical adequacy but for truth as well. Yet there are two distinct epistemic attitudes that can be taken: we can *accept* a theory (accept it as empirically adequate) or *believe* the theory (believe it to be true). We can take it to be the aim of science to produce a literally true story about the world, or simply to produce accounts that are empirically adequate. This is the issue of scientific realism versus its (divided) opposition. The intrascientific distinction between the observable and the unobservable is an anthropocentric distinction; but it is reasonable that the distinction should be drawn in terms of *us*, when it is a question of *our* attitudes toward theories.

NOTES

The author wishes to call attention to the improved and more extended treatment of this subject in his book *The Scientific Image* (Oxford: Oxford University Press, 1980) and to his reply to critics in C. A. Hooker and P. Churchland, eds., *Scientific Realism Versus Constructive Empiricism* (Chicago: University of Chicago Press, forthcoming).

1. For some criticisms, see my "Theoretical Entities: The Five Ways," *Philosophia* 4 (1974): 95-109, and "Wilfrid Sellars on Scientific Realism," *Dialogue* 14, 4 (December 1975): 606-616.

2. Cf. my "The Only Necessity Is Verbal Necessity," *Journal of Philosophy* 74, 2 (February 1977).

3. F. Cajori, ed., *Sir Isaac Newton's Mathematical Principles of Natural Philosophy and His System of the World* (Berkeley and Los Angeles: University of California Press, 1960), 12.

4. Herbert A. Simon, "The Axiomatization of Classical Mechanics," *Philosophy of Science* 21, 4 (October 1954): 340-343.

5. See, for example, Richard N. Boyd, "Realism, Undetermination, and a Causal Theory of Evidence," *Nous* 7, 1 (March 1973): 1-12.

6. Henri Poincaré, *The Value of Science*, trans. B. Halsted (New York: Dover, 1958), 98.

7. *Historical Development of Logic*, trans. J. Rosenthal (New York: Holt, 1929), 230.

8. B. Ellis, "The Origins and Nature of Newton's Laws of Motion," in *Beyond the Edge of Certainty*, ed. R. Colodny (Englewood Cliffs, N.J.: Prentice-Hall, 1965), 29-68.

9. For example, David Lewis, "How to Define Theoretical Terms," *Journal of Philosophy*, 67, 13 (July 9, 1970): 427-446. This paper is not subject to my criticisms here; on the contrary, it provides independent reasons to conclude that the empirical import of a theory cannot be syntactically isolated.

10. See n. 5. We could say that Boyd's paper, like Lewis's, provides independent evidence that empirical import cannot be syntactically isolated. But Boyd concludes also that there can be no distinction between truth and empirical adequacy for scientific theories.

11. C. Glymour, "Cosmology, Convention, and the Closed Universe," *Synthese* 24, nos. 1, 2 (July/August 1972): 195-218; discussed in my "Earman on the Causal Theory of Time," 87-95 (referred to therein by an earlier title).

12. See, for example, N. D. Cartwright, "Superposition and Macroscopic Observation," *Synthese* 29 (December 1974): 229-242, and references therein.

13. See Stanley Gudder, "Hidden Variables in Quantum Mechanics Reconsidered," *Review of Modern Physics* 40 (1968): 229-231; and section 3 of my "Semantic Analysis of Quantum Logic," in *Contemporary Research in the Foundations and Philosophy of Quantum Theory*, ed. C. A. Hooker (Dordrecht: D. Reidel, 1973), 80-113.

Index of Names

Index of Subjects

Abduction (inference to the best explanation), 3-5, 7, 16, 49, 65 f., 77, 85, 86, 103, 104, 220, 242
Ad hoc hypotheses, 30, 31, 120, 243
Auxiliary hypotheses, 50, 51, 57, 208 f.

Bayes's theorem, 31

Causality, 19, 22, 23
Constructivism, 19, 51 f., 55 f., 59, 60, 79, 89
Copernican theory, 175, 177, 178, 180, 184-187, 236
Crucial experiments, 53, 117
Cumulativity (of scientific knowledge), 199, 200, 216, 234 f., 244

Empirical equivalence (of theories), 42-44, 50, 51, 60, 61, 251-254, 257, 258
Empiricism, 18 f., 42, 45, 47, 55 f., 60, 67, 74, 79
Explanation, 125, 127, 128, 131, 188, 241, 242; structural, 26, 33, 34, 40; theories of, 178 f.

Fertility of theories, 30, 31

Idealism, 13, 16, 140, 159, 169, 170

Incommensurability (of theories), 52-56, 99, 144, 157, 194, 204; of modes of discourse, 156
Instrumentalism, 9, 24, 25, 36, 39, 74, 84, 87, 89, 92, 95, 114, 124, 125, 130, 131, 134, 159, 169, 196, 212, 231, 241, 243

Materialism, 24, 39
Michelson-Morely experimenta, 116, 117, 129

Newtonian theory, 9; gravitation, 11, 115, 118, 170, 176, 186, 205, 206 f.; mechanics, 10, 11, 118, 251 f.
Nominalism, 16, 20, 34, 159
Novel prediction, 30, 84, 88, 100, 243

Observation, 20, 21, 33, 48, 160, 168, 174, 175, 187, 198, 206, 212, 256
Observational-theoretical distinction, 44-47, 204, 255, 258
Occam's razor, 20, 109
Operationalism, 58, 62

Phenomenalism, 46, 74, 75, 161
Positivism, 9, 16, 19, 46, 55, 56, 59, 78, 91, 92, 108, 143, 161, 169, 217, 226, 237
Pragmatism, 23, 242

Designer: UC Press Staff
Compositor: Janet Sheila Brown
Printer: McNaughton & Gann, Inc.
Binder: John H. Dekker & Sons
Text: Paladium 10/12
Display: Paladium Bold